T0224372

Forum for Interdisciplinary Mathematics

Volume 2

Editor-in-chief

P.V. Subrahmanyam, Indian Institute of Technology Madras, Chennai, India

Editorial Board

Bhu Dev Sharma, Forum for Interdisciplinary Mathematics, Meerut, India
Janusz Matkowski, University of Zielona Góra, Zielona Góra, Poland
Mathieu Dutour Sikirić, Institute Rudjer Boúsković, Zagreb, Croatia
Thiruvenkatachari Parthasarathy, Chennai Mathematical Institute, Kelambakkam, India
Yogendra Prasad Chaubey, Concordia University, Montréal, Canada

The Forum for Interdisciplinary Mathematics (FIM) series publishes high-quality monographs and lecture notes in mathematics and interdisciplinary areas where mathematics has a fundamental role, such as statistics, operations research, computer science, financial mathematics, industrial mathematics, and bio-mathematics. It reflects the increasing demand of researchers working at the interface between mathematics and other scientific disciplines.

More information about this series at http://www.springer.com/series/13386

P.N. Natarajan

An Introduction to Ultrametric Summability Theory

Second Edition

Springer

P.N. Natarajan
Formerly of the Department of Mathematics
Ramakrishna Mission Vivekananda College
Chennai, Tamil Nadu
India

ISSN 2364-6748   ISSN 2364-6756   (electronic)
Forum for Interdisciplinary Mathematics
ISBN 978-81-322-3443-2   ISBN 978-81-322-2559-1   (eBook)
DOI 10.1007/978-81-322-2559-1

Springer New Delhi Heidelberg New York Dordrecht London
© Springer India 2014, 2015
Softcover reprint of the hardcover 2nd edition 2015

Printed on acid-free paper

Springer (India) Pvt. Ltd. is part of Springer Science+Business Media
(www.springer.com)

Dedicated to my parents S. Thailambal and P.S. Narayanasubramanian

Preface to the Second Edition

Most of the material discussed in the present revised, enlarged edition has appeared in the first edition of the book, *An Introduction to Ultrametric Summability Theory* (Springer, 2013). The first three chapters of the first edition have been retained in the second edition. In Chaps. 4–9 of the present edition, we present a survey of the literature on "ultrametric summability theory". We have supplemented substantially to the survey in the current edition. Our survey starts with a paper by Andree and Petersen of 1956 (it is the earliest known paper on the topic).

In Chap. 4, the Silverman–Toeplitz theorem is proved by using the "sliding hump method". Schur's theorem and Steinhaus theorem also find a mention here. It is proved that certain Steinhaus-type theorems fail to hold. An interesting characterization of infinite matrices in $(\ell_\alpha, \ell_\alpha)$, $\alpha > 0$ is proved. There is, as such, no classical analogue for this result. The core of a sequence and Knopp's core theorem are discussed. Towards the end of the chapter, an important result on Cauchy multiplication of series is proved. In Chap. 5, we introduce the Nörlund and Weighted mean methods in the ultrametric set-up, and their properties are elaborately discussed. We also show that the Mazur–Orlicz theorem and Brudno's theorem fail to hold in the ultrametric case. In Chap. 6, we introduce the Euler and Taylor methods and discuss their properties extensively. In Chap. 7, we prove Tauberian theorems for the Nörlund, the weighted mean and the Euler methods. In Chap. 8, we introduce double sequences and double series in ultrametric analysis. We prove Silverman–Toeplitz theorem for four-dimensional infinite matrices. We also prove Schur's and Steinhaus theorems for four-dimensional infinite matrices. Towards the end of the chapter, we obtain some interesting characterizations of two-dimensional Schur matrices. Finally, in Chap. 9, we introduce the Nörlund and the weighted mean methods for double sequences and make a detailed study of their properties. The author thanks E. Boopal for typing the manuscript.

Errors in the first edition, both typographical and conceptual, have been corrected in the present revised, enlarged edition.

Chennai, India   P.N. Natarajan

Preface to the First Edition

The purpose of the present monograph is to discuss briefly what summability theory is like when the underlying field is not $\mathbb{R}$ (the field of real numbers) or $\mathbb{C}$ (the field of complex numbers) but a field K with a non-Archimedean or ultrametric valuation, i.e., a mapping $|\cdot|: K \to \mathbb{R}$ satisfying the ultrametric inequality $|x + y| \leq \max(|x|, |y|)$ instead of the usual triangle inequality $|x + y| \leq |x| + |y|$, $x, y \in K$.

To make the monograph really useful to those who wish to take up the study of ultrametric summability theory and do some original work therein, some knowledge of real and complex analysis, functional analysis and summability theory over $\mathbb{R}$ or $\mathbb{C}$ is assumed.

Some of the basic properties of ultrametric fields—their topological structure and geometry—are discussed in Chap. 1. In this chapter, we introduce the p-adic valuation, p being prime and prove that any valuation of $\mathbb{Q}$ (the field of rational numbers) is either the trivial valuation, a p-adic valuation or a power of the usual absolute value $|\cdot|_\infty$ on $\mathbb{R}$, i.e., $|\cdot|_\infty^\alpha$, where $0 < \alpha \leq 1$. We discuss equivalent valuations too. In Chap. 2, we discuss some arithmetic and analysis in $\mathbb{Q}_p$, the p-adic field for a prime p. In Chap. 2, we also introduce the concepts of differentiability and derivatives in ultrametric analysis and very briefly indicate how ultrametric calculus is different from our usual calculus.

In Chap. 3, we speak of ultrametric Banach space, and also mention the many results of the classical Banach space theory, viz., the closed graph, the open mapping and the Banach-Steinhaus theorems carry over in the ultrametric set-up. However, the Hahn-Banach theorem fails to hold. To salvage the Hahn-Banach theorem, the concept of a "spherically complete field" is introduced and Ingleton's version of the Hahn-Banach theorem is proved. The lack of ordering in an ultrametric field K makes it quite difficult to find a substitute for classical "convexity". However, classical convexity is replaced, in the ultrametric setting, by a notion called "K-convexity", which is briefly discussed towards the end of the chapter.

In the main Chap. 4, our survey of the literature on "Ultrametric Summability Theory", starts with the paper of Andree and Petersen of 1956 (it was the earliest known paper on the topic) to the present. As far as the author of the present monograph knows, most of the material discussed in the survey has not appeared in book form earlier. Almost all of Chap. 4 consists entirely of the work of the author of the present monograph. Suitable references have been provided at appropriate places indicating where further developments may be found.

The author takes this opportunity to thank Prof. P. V. Subrahmanyam of the Department of Mathematics, Indian Institute of Technology (Madras), Chennai for encouraging him to write the monograph during the author's short stay at the Institute (July 8–August 5, 2011) as a Visiting Faculty. The author thanks the Forum for Inter-disciplinary Mathematics for being instrumental in getting the monograph published.

Chennai, India   P.N. Natarajan

Contents

About the Author

P.N. Natarajan is former professor and head, Department of Mathematics, Ramakrishna Mission Vivekananda College, Chennai. He did his Ph.D. from the University of Madras, under Prof. M.S. Rangachari, former director and head, The Ramanujan Institute for Advanced Study in Mathematics, University of Madras. An active researcher, Prof. Natarajan has over 100 research papers to his credit published in several international journals like *Proceedings of the American Mathematical Society, Bulletin of the London Mathematical Society, Indagationes Mathematicae, Annales Mathematiques Blaise Pascal*, and *Commentationes Mathematicae* (Prace Matematyczne). His research interest includes summability theory and functional analysis (both classical and ultrametric). Professor Natarajan was honoured with the Dr. Radhakrishnan Award for the Best Teacher in Mathematics for the year 1990–1991 by the Government of Tamil Nadu. Besides visiting several institutes of repute in Canada, France, Holland and Greece on invitation, Prof. Natarajan has participated in several international conferences and has chaired sessions.

Chapter 1
Introduction and Preliminaries

Abstract Some basic properties of ultrametric fields—their topological structure and geometry are discussed in this chapter. We introduce the p-adic valuation, p being prime and prove that any valuation of $\mathbb{Q}$ (the field of rational numbers) is either the trivial valuation, a p-adic valuation or a power of the usual absolute value, where the power is positive and less than or equal to 1. We discuss equivalent valuations too.

Keywords Archimedean axiom · Ultrametric valuation · Ultrametric field · p-adic valuation · p-adic field · p-adic numbers · Equivalent valuations

The purpose of this book is to introduce a "NEW ANALYSIS" to students of Mathematics at the undergraduate and post graduate levels, which in turn introduces a geometry very different from the familiar Euclidean geometry and Riemannian geometry. Strange things happen: for instance, 'every triangle is isosceles' and 'every point of a sphere is a centre of the sphere'!.

'Analysis' is that branch of Mathematics where we use the idea of limits extensively. A study of Analysis starts with limits, continuity, differentiability etc. and almost all mathematical models are governed by differential equations over the field $\mathbb{R}$ of real numbers. $\mathbb{R}$ has a geometry which is Euclidean. Imagine a pygmy tortoise trying to travel along a very long path; assume that its destination is at a very long distance from its starting point. If at every step, the pygmy tortoise covers a small distance ϵ, can it ever reach its destination, assuming that the tortoise has infinite life? Our common sense says "Yes". It is one of the important axioms in Euclidean geometry that "Any large segment on a straight line can be covered by successive addition of small segments along the same line". It is equivalent to the statement: "given any number $M > 0$, there exists an integer N such that $N > M$". This is familiarly known as the "Archimedean axiom" of the real number field $\mathbb{R}$. What would happen if we do not have this axiom? Are there fields which are non-archimedean? In the sequel, we will show that such fields exist and the metric on such fields introduces a geometry very different from the familiar Euclidean geometry and Riemannian geometry. In such a geometry, every triangle is isoceles; given two spheres, either

© Springer India 2015

P.N. Natarajan, *An Introduction to Ultrametric Summability Theory*,
Forum for Interdisciplinary Mathematics 2, DOI 10.1007/978-81-322-2559-1_1

they are disjoint or identical and consequently every point of a sphere is a centre of the sphere!

Let us now revisit the field $\mathbb{Q}$ of rational numbers. The usual absolute value $|\cdot|$ on $\mathbb{Q}$ is defined as

$$|x| = \begin{cases} x, & \text{if } x \geq 0; \\ -x, & \text{if } x < 0. \end{cases}$$

We recall the process by which we get real numbers from rational numbers. Any real number x has a decimal expansion and we can write

$$x = \pm 10^{\beta}(x_0 + x_1 10^{-1} + x_2 10^{-2} + \cdots), \quad \beta \in \mathbb{Z}, \tag{*}$$
$$x_j = 0, 1, \ldots, 9, j = 0, 1, 2, \ldots, \mathbb{Z} \text{ being the ring of integers.}$$

We note that the decimal expansion of a real number x enables us to look at x as the limit of a sequence of rational numbers. The metric space $(\mathbb{Q}, |\cdot|)$, where $|\cdot|$ is the usual metric on $\mathbb{Q}$, is not complete in the sense that there are Cauchy sequences in $\mathbb{Q}$ which do not converge in $\mathbb{Q}$. We now consider the set ζ of all Cauchy sequences in $\mathbb{Q}$ which do not have a limit in $\mathbb{Q}$. Declare that any two sequences $\{x_n\}$ and $\{y_n\}$ are equivalent if $\{x_n - y_n\}$ converges to 0 in $(\mathbb{Q}, |\cdot|)$. After such an identification, the set of all equivalence classes in ζ is precisely the set $\mathbb{R}$ of all real numbers. Note that $\mathbb{Q}$ is dense in $\mathbb{R}$. The metric d in $\mathbb{R}$ is defined as follows:

$$d(x, y) = \lim_{n \to \infty} |x_n - y_n|,$$

where $\{x_n\}, \{y_n\}$ are the Cauchy sequences of rational numbers representing the real numbers x and y respectively. This procedure of embedding $\mathbb{Q}$, as a dense subspace, in a complete metric space $\mathbb{R}$ is called the completion of $\mathbb{Q}$. We can imitate this process and embed any given metric space, as a dense subspace, in a complete metric space and speak of the "completion of a metric space".

Given a field K, by a valuation of K, we mean a mapping $|\cdot| : K \to \mathbb{R}$ such that

$$|x| \geq 0; \quad |x| = 0 \text{ if and only if } x = 0; \tag{1.1}$$

$$|xy| = |x||y|; \tag{1.2}$$

and

$$|x + y| \leq |x| + |y|, \tag{1.3}$$

$x, y \in K$. The field K, with the valuation $|\cdot|$, is called a valued field.

Example 1.1 The usual absolute value function is a valuation of $\mathbb{Q}$.

Example 1.2 Given any field K, define a valuation $|\cdot|$ of K as follows:

$$|x| = \begin{cases} 0, & \text{if } x = 0; \\ 1, & \text{if } x \neq 0. \end{cases}$$

In Example 1.2 above, $|\cdot|$ is called the trivial valuation of K. Any valuation of K which is not the trivial valuation is called a non-trivial valuation. If K is a valued field with valuation $|\cdot|$, define

$$d(x, y) = |x - y|, \tag{1.4}$$

$x, y \in K$. It is easy to check that d is a metric on K so that K is a metric space with respect to the 'metric d induced by the valuation' defined by (1.4). Consequently, we can define topological concepts like open set, closed set, convergence, etc. in valued fields.

We now deal with $\mathbb{Q}$ in a different direction.

Definition 1.1 Let c be any fixed real number such that $0 < c < 1$ and p be a fixed prime number. We define a valuation, denoted by $|\cdot|_p$, of $\mathbb{Q}$ as follows: Define $|0|_p = 0$; if $x \in \mathbb{Q}$, $x \neq 0$, we write x in the form

$$x = p^\alpha \left(\frac{a}{b} \right),$$

where $\alpha, a \in \mathbb{Z}, b \in \mathbb{Z}^+ = \{1, 2, \ldots\}$, p does not divide a, b and $(a, b) = 1$. Noting that the above form of x is unique, define $|x|_p = c^\alpha$.

It is now worthwhile to check that $|\cdot|_p$ is a valuation of $\mathbb{Q}$. From the definition, it is clear that $|x|_p \geq 0$ and $|x|_p = 0$ if and only if $x = 0$. For $y \in \mathbb{Q}$, we shall write

$$y = p^\beta \left(\frac{a'}{b'} \right),$$

where $\beta, a' \in \mathbb{Z}, b' \in \mathbb{Z}^+$, p does not divide a', b' and $(a', b') = 1$ so that $|y|_p = c^\beta$. We now have,

$$xy = p^{\alpha+\beta} \left(\frac{aa'}{bb'} \right),$$

where $\alpha + \beta, aa' \in \mathbb{Z}, bb' \in \mathbb{Z}^+$, p does not divide aa', bb' and $(aa', bb') = 1$, this being so since p is a prime number (this is the reason why p was chosen to be a prime number!). Thus

$$|xy|_p = c^{\alpha+\beta} = c^\alpha \cdot c^\beta = |x|_p |y|_p.$$

We now claim that

$$|x + y|_p \leq \max(|x|_p, |y|_p), \tag{1.5}$$

which is stronger than (1.3). To prove (1.5), we show that

$$|x|_p \leq 1 \Rightarrow |1+x|_p \leq 1. \tag{1.6}$$

Leaving out the trivial case, let $x \neq 0$. It follows from the definition that $|x|_p \leq 1 \Rightarrow$ $\alpha \geq 0$ and x can be written in the form

$$x = \frac{c'}{d'},$$

where c', d' are integers which are relatively prime and p does not divide d'. Now,

$$1 + x = 1 + \frac{c'}{d'} = \frac{c'+d'}{d'}.$$

Since $1 + x$ has a denominator which is prime to p, we have

$$|1+x|_p \leq 1,$$

proving our claim. Now, if $y = 0$, (1.5) is trivially true. Let $y \neq 0$. Without loss of generality, we shall suppose that $|x|_p \leq |y|_p$ so that $\left|\dfrac{x}{y}\right|_p \leq 1$. Using (1.6), we have, $\left|1 + \dfrac{x}{y}\right|_p \leq 1$ and so

$$\begin{aligned}|x+y|_p = \left|y\left(1+\frac{x}{y}\right)\right|_p &= |y|_p \left|1+\frac{x}{y}\right|_p \\ &\leq |y|_p \\ &= \max(|x|_p, |y|_p),\end{aligned}$$

establishing (1.5).

If a valuation of K satisfies (1.5) too, it is called a 'non-archimedean valuation' of K and K, with such a valuation, is called a 'non-archimedean valued field' or just a 'non-archimedean field'. The metric induced by a non-archimedean valuation satisfies the much stronger inequality

$$d(x,y) \leq \max(d(x,z), d(z,y)), \tag{1.7}$$

which is known as the 'ultrametric inequality'. Any metric which satisfies (1.7) is known as an 'ultrametric'. Study of analysis in non-archimedean fields is known as 'non-archimedean analysis' or 'p-adic analysis' or 'ultrametric analysis'. A non-archimedean valuation is called an 'ultrametric valuation' and a non-archimedean field is called an 'ultrametric field'.

We have thus proved that $|\cdot|_p$ is a non-archimedean valuation of the rational number field $\mathbb{Q}$. The completion of $\mathbb{Q}$ with respect to the p-adic valuation $|\cdot|_p$ is called the p-adic field, denoted by $\mathbb{Q}_p$. Elements of $\mathbb{Q}_p$ are called p-adic numbers.

We are very familiar with analysis in $\mathbb{R}$ (i.e., real analysis) or in $\mathbb{C}$ (i.e., complex analysis). We now prove some elementary results in ultrametric analysis. These results point out significant departures from real or complex analysis. In the sequel we shall suppose that K is a non-archimedean field with valuation $|\cdot|$.

Theorem 1.1 *If K is a non-archimedean field and if $|a| \neq |b|$, $a, b \in K$, then*

$$|a + b| = \max(|a|, |b|).$$

Proof For definiteness, let us suppose that $|a| > |b|$. Now,

$$|a| = |(a + b) - b| \leq \max(|a + b|, |b|)$$
$$= |a + b|,$$

for, otherwise, $|a| \leq |b|$, a contradiction of our assumption. Consequently,

$$|a| \leq |a + b| \leq \max(|a|, |b|) = |a|,$$

and so

$$|a + b| = \max(|a|, |b|),$$

completing the proof.   □

Corollary 1.1 *Any triangle is isosceles in the following sense: given a triad of points in an ultrametric field, at least two pairs of points have the same distance.*

Proof Let d be the ultrametric induced by the non-archimedean valuation $|\cdot|$ of K. Consider a triangle with vertices x, y and z.

Cases 1 If the triangle is equilateral, i.e., $d(x, y) = d(y, z) = d(z, x)$, the triangle is already isosceles and there is nothing to prove!

Cases 2 Suppose the triangle is not equilateral (say) $d(x, y) \neq d(y, z)$, i.e., $|x - y| \neq |y - z|$. Then

$$d(x, z) = |x - z| = |(x - y) + (y - z)|$$
$$= \max(|x - y|, |y - z|) \text{ using Theorem 1.1}$$
$$= \max(d(x, y), d(y, z)),$$

proving our claim.   □

Theorem 1.2 *Every point of the open sphere*

$$S_\epsilon(x) = \{y \in K / |y - x| < \epsilon\}, \quad \epsilon > 0$$

is a centre, i.e., if $y \in S_\epsilon(x)$, $S_\epsilon(y) = S_\epsilon(x)$.

Proof Let $y \in S_\epsilon(x)$ and $z \in S_\epsilon(y)$. Then $|y - x| < \epsilon$ and $|z - y| < \epsilon$ so that

$$
\begin{aligned}
|z - x| &= |(z - y) + (y - x)| \\
&\leq \max(|z - y|, |y - x|) \\
&< \epsilon.
\end{aligned}
$$

Thus $z \in S_\epsilon(x)$. Consequently $S_\epsilon(y) \subseteq S_\epsilon(x)$. The reverse inclusion is similarly proved so that $S_\epsilon(y) = S_\epsilon(x)$, completing the proof. $\qquad\square$

Corollary 1.2 *Any two spheres are either disjoint or one is contained in the other. More specifically, given two spheres $S_{\epsilon_1}(x)$, $S_{\epsilon_2}(y)$ with $\epsilon_1 \leq \epsilon_2$, either $S_{\epsilon_1}(x) \cap S_{\epsilon_2}(y) = \emptyset$ or $S_{\epsilon_1}(x) \subseteq S_{\epsilon_2}(y)$.*

Theorem 1.3 *Any sequence $\{x_n\}$ in K is a Cauchy sequence if and only if*

$$
|x_{n+1} - x_n| \to 0, \quad n \to \infty. \tag{1.8}
$$

Proof If $\{x_n\}$ is a Cauchy sequence in K, it is clear that (1.8) holds. Conversely, let (1.8) hold. So given $\epsilon > 0$, there exists a positive integer N such that

$$
|x_{n+1} - x_n| < \epsilon, \quad n \geq N. \tag{1.9}
$$

Now, for definiteness, let $m > n$. Then

$$
\begin{aligned}
|x_m - x_n| &= |(x_m - x_{m-1}) + (x_{m-1} - x_{m-2}) + \cdots + (x_{n+1} - x_n)| \\
&\leq \max(|x_m - x_{m-1}|, |x_{m-1} - x_{m-2}|, \ldots, |x_{n+1} - x_n|) \\
&< \epsilon,
\end{aligned}
$$

$m > n \geq N$, in view of (1.9). This proves that $\{x_n\}$ is a Cauchy sequence. $\qquad\square$

In classical analysis i.e., real or complex analysis, we know that if $\sum_{n=1}^{\infty} x_n$ converges, then $x_n \to 0, n \to \infty$. However, the converse is not true. For instance, $\sum_{n=1}^{\infty} \frac{1}{n}$ diverges, though $\frac{1}{n} \to 0, n \to \infty$. We will now prove that the converse is also true in the case of complete, non-archimedean fields.

Theorem 1.4 *Let K be a complete, non-archimedean field. Then $\sum_{n=1}^{\infty} a_n, a_n \in K$, $n = 1, 2, \ldots$, converges if and only if*

$$
a_n \to 0, \quad n \to \infty.
$$

Proof Let $\sum_{n=1}^{\infty} a_n$ converge and $s_n = \sum_{k=1}^{n} a_k, n = 1, 2, \ldots$. Then $\{s_n\}$ converges and so

$$|a_n| = |s_n - s_{n-1}| \to 0, \quad n \to \infty.$$

Thus $a_n \to 0, n \to \infty$. Conversely, let $a_n \to 0, n \to \infty$. Now,

$$|s_n - s_{n-1}| = |a_n| \to 0, \quad n \to \infty.$$

In view of Theorem 1.3, $\{s_n\}$ is a Cauchy sequence. Since K is complete with respect to the valuation, $\lim_{n \to \infty} s_n$ exists in K and so $\sum_{n=1}^{\infty} a_n$ converges. $\square$

Going deeper and deeper into ultrametric analysis, one can point out more and more results, which indicate significant departure from classical analysis—in fact ultrametric analysis is a "STRANGE AND RUGGED PATH"!. One is tempted to ask: "Are there natural non-archimedean phenomena?". We have observed earlier, as a corollary of Theorem 1.2, that any two spheres in a non-archimedean field K are either disjoint or one is contained in the other. This situation is amply illustrated by two drops of mercury on a smooth surface, which is a typical non-archimedean phenomenon. When we face with such situations, ultrametric analysis would be appropriate.

Bachman's book [1] is an excellent intoduction to non-archimedean analysis.

If K is a field with an ultrametric valuation, then, for any integer n,

$$|n| \leq \max(\underbrace{|1|, |1|, \ldots, |1|}_{n \text{ times}})$$

$$= |1| = 1$$

(note that $|1| = |1.1| = |1||1|$ implying that $|1| = 1$ using the fact that $|1| \neq 0$).

Thus $\{n\}_{n=1}^{\infty}$ is a bounded sequence (note that this is the reason why $| \cdot |$ is non-archimedean and the field K is a non-archimedean field!). Converse is also true. This fact is established by the following result.

Theorem 1.5 *Let K be a field with a valuation $| \cdot |$ and let there exist $d > 0$ such that $|m| \leq d$ for all integers m of K, where the set of integers refers to the isomorphic image of $\mathbb{Z}$ in K, if K has characteristic 0 or the isomorphic image of the residue classes modulo p, if K has characteristic p. Then the valuation $| \cdot |$ is non-archimedean.*

Proof If $a, b \in K$ and τ is a positive integer,

$$|a + b|^{\tau} = |(a + b)^{\tau}|$$

$$= \left| a^{\tau} + \binom{\tau}{1} a^{\tau-1} b + \binom{\tau}{2} a^{\tau-2} b^2 + \cdots + b^{\tau} \right|$$

$$\leq |a|^\tau + \left|\binom{\tau}{1}\right| |a|^{\tau-1}|b| + \cdots + |b|^\tau$$

$$\leq (\tau + 1)d \max(|a|, |b|)^\tau.$$

Taking τth roots on both sides and allowing $\tau \to \infty$, we have,

$$|a + b| \leq \max(|a|, |b|),$$

proving that the valuation $|\cdot|$ is non-archimedean. □

We already obtained several valuations of $\mathbb{Q}$, viz., the p-adic valuations, the usual absolute value and the trivial valuation. We will prove that these are the only valuations of $\mathbb{Q}$ in a certain sense. So we wish to find all the valuations of $\mathbb{Q}$. To achieve this, we need the following lemmas.

Lemma 1.1 *If $0 < r \leq s$ and a_i, $i = 1, 2, \ldots, n$ are non-negative real numbers, then*

$$\left(\sum_{i=1}^n a_k^s\right)^{\frac{1}{s}} \leq \left(\sum_{k=1}^n a_k^r\right)^{\frac{1}{r}}.$$

Proof Let $d = \sum_{k=1}^n a_k^r$. Then

$$\frac{\left(\sum_{k=1}^n a_k^s\right)^{\frac{1}{s}}}{d^{\frac{1}{r}}} = \left(\sum_{k=1}^n \frac{a_k^s}{d^{\frac{s}{r}}}\right)^{\frac{1}{s}}$$

$$= \left\{\sum_{k=1}^n \left(\frac{a_k^r}{d}\right)^{\frac{s}{r}}\right\}^{\frac{1}{s}}$$

$$\leq \left\{\sum_{k=1}^n \frac{a_k^r}{d}\right\}^{\frac{1}{s}},$$

since $\left(\frac{a_k^r}{d}\right)^{\frac{s}{r}} \leq \frac{a_k^r}{d}$, this being so since $\frac{a_k^r}{d} \leq 1$ and $\frac{s}{r} \geq 1$. However,

$$\sum_{k=1}^n \frac{a_k^r}{d} = \frac{1}{d}\sum_{k=1}^n a_k^r = \frac{1}{d}d = 1,$$

so that

$$\frac{\left(\sum_{k=1}^{n} a_k^s\right)^{\frac{1}{s}}}{d^{\frac{1}{r}}} \leq 1,$$

i.e., $\left(\sum_{k=1}^{n} a_k^s\right)^{\frac{1}{s}} \leq d^{\frac{1}{r}}$

$$= \left(\sum_{k=1}^{n} a_k^r\right)^{\frac{1}{r}},$$

completing the proof. $\qquad\square$

Lemma 1.2 *If* $0 < \alpha \leq 1$ *and* b_i, $i = 1, 2, \ldots, n$ *are non-negative real numbers, then*

$$(b_1 + b_2 + \cdots + b_n)^{\alpha} \leq b_1^{\alpha} + b_2^{\alpha} + \cdots + b_n^{\alpha}.$$

Proof In Lemma 1.1, choose $a_i = b_i$, $i = 1, 2, \ldots, n$ and $s = 1, r = \alpha$. Then, we have

$$b_1 + b_2 + \cdots + b_n \leq (b_1^{\alpha} + b_2^{\alpha} + \cdots + b_n^{\alpha})^{\frac{1}{\alpha}},$$
$$\text{i.e., } (b_1 + b_2 + \cdots + b_n)^{\alpha} \leq b_1^{\alpha} + b_2^{\alpha} + \cdots + b_n^{\alpha}. \qquad\square$$

Going back to our problem, let m, n be integers $> 1, m \geq n$. Writing m using the base n, we have,

$$m = a_0 + a_1 n + a_2 n^2 + \cdots + a_k n^k,$$

where $0 \leq a_i \leq n - 1, i = 0, 1, 2, \ldots, k, n^k \leq m$. So,

$$k \leq \frac{\log m}{\log n}. \qquad (1.10)$$

Let $|\cdot|$ be a valuation of $\mathbb{Q}$. Now, note that

$$|a_i| < n, \quad i = 0, 1, 2, \ldots, k$$

so that

$$|m| \leq |a_0| + |a_1||n| + |a_2||n|^2 + \cdots + |a_k||n|^k$$
$$< n(1 + |n| + |n|^2 + \cdots + |n|^k)$$
$$\leq n(k + 1) \max(1, |n|)^k$$

$$\leq n \left(\frac{\log m}{\log n} + 1 \right) \max(1, |n|)^{\frac{\log m}{\log n}}, \ \text{using (1.10).} \tag{1.11}$$

Since (1.11) holds for any integers $m, n > 1$, replace m by m^τ, $\tau \in \mathbb{Z}^+$ in (1.11). Thus

$$|m^\tau| \leq n \left(\frac{\log m^\tau}{\log n} + 1 \right) \max(1, |n|)^{\frac{\log m^\tau}{\log n}},$$

$$\text{i.e., } |m|^\tau \leq n \left(\frac{\tau \log m}{\log n} + 1 \right) \max(1, |n|)^{\frac{\tau \log m}{\log n}}.$$

Taking τth root on both sides,

$$|m| \leq \left\{ n \left(\frac{\tau \log m}{\log n} + 1 \right) \right\}^{\frac{1}{\tau}} \max(1, |n|)^{\frac{\log m}{\log n}}.$$

Allowing $\tau \to \infty$ and using the fact that $\lim_{\tau \to \infty} \tau^{\frac{1}{\tau}} = 1$, we have,

$$|m| \leq \max(1, |n|)^{\frac{\log m}{\log n}}. \tag{1.12}$$

Cases 1 There exists $n > 1$ such that $|n| \leq 1$. Now (1.12) implies that $|m| \leq 1$ for all $m > 1$. Using $|-1| = 1$, we see that $|m| \leq 1$ for all $m \in \mathbb{Z}$. In view of Theorem 1.5, it follows that the valuation in this case is non-archimedean. If $|m| = 1$ for all $m \in \mathbb{Z}$, $m \neq 0$, it follows that $|\cdot|$ is the trivial valuation. On the other hand, let us suppose that $|m| < 1$ for some $m \in \mathbb{Z}$, $m > 1$. Let p be the least positive integer such that $|p| < 1$. Then p must be a prime. For, otherwise, $p = ab$, where a, b are integers > 1, $a, b < p$. Then,

$$|p| = |ab| = |a||b| = 1.1 = 1,$$

which is a contradiction. Let

$$m = qp + r, \ 0 \leq r < p.$$

We claim that $r = 0$. For, otherwise, i.e., if $r \neq 0$, since $r < p$, $|r| = 1$. But

$$|qp| = |\underbrace{p + p + \cdots + p}_{q \text{ times}}| \leq |p| < 1,$$

so that

$$|m| = \max(|qp|, |r|) = 1, \text{ in view of Theorem 1.1}$$

which is a contradiction. Thus $r = 0$ and so p/m. Conversely, if p/m, then $|m| < 1$. Consequently, $|m| < 1$ if and only if p/m. If $x \in \mathbb{Q}, x \neq 0$, then x can be written as

$$x = p^\alpha \frac{a}{b},$$

where $a, b, \alpha \in \mathbb{Z}, b \neq 0, p \nmid a, p \nmid b$. In view of the above discussion,

$$|x| = |p|^\alpha \frac{|a|}{|b|} = |p|^\alpha, \text{ since } |a| = |b| = 1$$

$$= c^\alpha,$$

where $c = |p|$ is some real number such that $0 < c < 1$. Thus in this case, $|\cdot|$ is the p-adic valuation.

Cases 2   For any integer $n > 1, |n| > 1$. From (1.12) we have,

$$|m|^{\frac{1}{\log m}} \leq |n|^{\frac{1}{\log n}}.$$

Since this is true for all integers $m, n > 1$, interchanging m and n, we obtain

$$|m|^{\frac{1}{\log m}} \geq |n|^{\frac{1}{\log n}}.$$

Thus

$$|m|^{\frac{1}{\log m}} = |n|^{\frac{1}{\log n}},$$

for all integers $m, n > 1$. In other words,

$$|m|^{\frac{1}{\log m}} = c,$$

where c is independent of m. Noting that $c > 1$, we can write

$$c = e^\alpha, \text{ for some } \alpha > 0.$$

$$\text{i.e., } |m|^{\frac{1}{\log m}} = e^\alpha$$

$$\text{i.e., } |m| = e^{\alpha \log m} = e^{\log m^\alpha} = m^\alpha. \tag{1.13}$$

In particular,

$$|2| = 2^\alpha$$

so that

$$2^\alpha = |2| = |1 + 1| \leq |1| + |1| = 1 + 1 = 2.$$

Consequently

$$\alpha \le 1.$$

For any integer $m > 1$, using (1.13),

$$|m| = m^\alpha = |m|_\infty^\alpha, \tag{1.14}$$

where $|\cdot|_\infty$ denotes the usual absolute value function on $\mathbb{R}$. Note that (1.14) holds trivially for $m = 0$. Since $|-1| = 1$, $|-1|_\infty = 1$, (1.14) is true for all $m \in \mathbb{Z}$. So for any $x \in \mathbb{Q}$,

$$|x| = |x|_\infty^\alpha, \quad \text{where } 0 < \alpha \le 1.$$

We shall show that all functions of the form $|\cdot|_\infty^\alpha$, $0 < \alpha \le 1$ are valuations of $\mathbb{Q}$. Clearly (1.1) and (1.2) are satisfied.

Now,

$$|x + y|_\infty^\alpha \le (|x|_\infty + |y|_\infty)^\alpha \le |x|_\infty^\alpha + |y|_\infty^\alpha,$$

since $0 < \alpha \le 1$, using Lemma 1.2. Thus, we have proved the following interesting result.

Theorem 1.6 *Any valuation of $\mathbb{Q}$ is either the trivial valuation, a p-adic valuation or a power of the usual absolute value, i.e., $|\cdot|_\infty^\alpha$, where $0 < \alpha \le 1$.*

We shall now prove that all valuations of the form $|\cdot|_\infty^\alpha$, $0 < \alpha \le 1$, are equivalent in a sense which will be made precise in the sequel.

Definition 1.2 Let $|\cdot|_1, |\cdot|_2$ be two non-trivial valuations of a field K. $|\cdot|_1$ is said to be equivalent to $|\cdot|_2$, written as $|\cdot|_1 \sim |\cdot|_2$, if $|a|_1 < 1$ implies $|a|_2 < 1$, $a \in K$.

Note that if every null sequence with respect to $|\cdot|_1$ is also a null sequence with respect to $|\cdot|_2$ then $|\cdot|_1 \sim |\cdot|_2$. Let $|a|_1 < 1$. Then $\{a^n\}$ is a null sequence with respect to $|\cdot|_1$. By hypothesis, $\{a^n\}$ is a null sequence with respect to $|\cdot|_2$ too and so we should have $|a|_2 < 1$. Thus $|\cdot|_1 \sim |\cdot|_2$.

We also note that if $|\cdot|_1 \sim |\cdot|_2$ and $|a|_1 > 1$, then $|a|_2 > 1$. For, if $|a|_1 > 1$, then $\left|\frac{1}{a}\right|_1 < 1$. Since $|\cdot|_1 \sim |\cdot|_2$, $\left|\frac{1}{a}\right|_2 < 1$ and so $|a|_2 > 1$.

Theorem 1.7 *If $|\cdot|_1 \sim |\cdot|_2$, then $|a|_1 = 1$ implies $|a|_2 = 1$.*

Proof Since $|\cdot|_1$ is non-trivial, there exists $b \in K$, $b \ne 0$ such that $|b|_1 < 1$. Now,

$$|a^n b|_1 = |a|_1^n |b_1| = |b|_1 < 1.$$

Since $|\cdot|_1 \sim |\cdot|_2$, $|a^n b|_2 < 1$,

$$\text{i.e., } |a|_2^n |b|_2 < 1,$$

$$\text{i.e., } |a|_2 < \left(\frac{1}{|b|_2}\right)^{\frac{1}{n}}.$$

Allowing $n \to \infty$, we get $|a|_2 \leq 1$. Replacing a by $\frac{1}{a}$, we get $\left|\frac{1}{a}\right|_2 \leq 1$, i.e., $|a|_2 \geq 1$. Hence $|a|_2 = 1$.

It is clear that $\sim$ is reflexive and transitive. Let $|\cdot|_1 \sim |\cdot|_2$ and $|a|_2 < 1$. We claim that $|a|_1 < 1$. For, if $|a|_1 = 1$, then $|a|_2 = 1$, using Theorem 1.7. Also if $|a|_1 > 1$, then $|a|_2 > 1$, using the comment preceding Theorem 1.7. In both the cases, we have a contradiction. So $|a|_1 < 1$. Thus $\sim$ is symmetric and consequently $\sim$ is an equivalence relation. □

Theorem 1.8 *If* $|\cdot|_1 \sim |\cdot|_2$, *then* $|\cdot|_2 = |\cdot|_1^{\gamma}$, *where* γ *is a positive real number.*

Proof Let $b \in K$ be fixed such that $|b|_1 > 1$. Let $a \in K$, $a \neq 0$.
Then

$$|a|_1 = |b|_1^{\alpha}, \quad \text{where } \alpha = \frac{\log |a|_1}{\log |b|_1}.$$

Let $n, m \in \mathbb{Z}$ such that $\frac{n}{m} > \alpha$. Then

$$|a|_1 < |b|_1^{\frac{n}{m}},$$

$$\text{i.e., } \left|\frac{a^m}{b^n}\right|_1 < 1.$$

Since by hypothesis, $|\cdot|_1 \sim |\cdot|_2$,

$$\left|\frac{a^m}{b^n}\right|_2 < 1,$$

from which we get

$$|a|_2 < |b|_2^{\frac{n}{m}}.$$

Similarly, if $\frac{n}{m} < \alpha$, we get,

$$|a|_2 > |b|_2^{\frac{n}{m}}.$$

Consequently, we have,

$$|a|_2 = |b|_2^{\alpha}.$$

Thus

$$\alpha = \frac{\log |a|_1}{\log |b|_1} = \frac{\log |a|_2}{\log |b|_2}$$

and so

$$\log |a|_2 = \frac{\log |b|_2}{\log |b|_1} \cdot \log |a|_1$$
$$= \gamma \log |a|_1,$$

where $\gamma = \dfrac{\log |b|_2}{\log |b|_1}$. Then

$$|a|_2 = |a|_1^{\gamma} \quad \text{(where } \gamma = \frac{\log |b|_2}{\log |b|_1} \text{ is a positive real number)}.$$

The above result is trivially true for $a = 0$. The proof of the theorem is now complete.   □

Theorem 1.8 yields an alternate characterization of equivalent valuations.

Theorem 1.9 *The non-trivial valuations* $| \cdot |_1, | \cdot |_2$ *are equivalent if and only if any null sequence with respect to* $| \cdot |_1$ *is also a null sequence with respect to* $| \cdot |_2$.

Proof If $| \cdot |_1 \sim | \cdot |_2$, in view of Theorem 1.8, any null sequence with respect to $| \cdot |_1$ is also a null sequence with respect to $| \cdot |_2$. By the comment immediately following Definition 1.2, the converse is also true.   □

Using Theorem 1.8, we can "reformulate" Theorem 1.6 as follows.

Theorem 1.10 *Any valuation of* $\mathbb{Q}$ *is either the trivial valuation, a p-adic valuation or the usual absolute value (upto equivalent valuations).*

In the context of Theorem 1.8, we note that the choice of c, $0 < c < 1$, used to define the p-adic valuation $| \cdot |_p$, has no impact on the study of the valuation $| \cdot |_p$ in the sense that for different values of c, $0 < c < 1$, the corresponding valuations $| \cdot |_p$ are equivalent.

It is worthwhile for the reader to try the following exercises.

Exercise 1.1 Prove that any valuation of a field of characteristic not equal to 0 is non-archimedean.

Exercise 1.2 Prove that the only valuation of a finite field is the trivial valuation.

Exercise 1.3 Prove that $\mathbb{Q}$ is not complete with respect to the p-adic valuation $| \cdot |_p$.

Reference

1. Bachman, G.: Introduction to p-adic Numbers and Valuation Theory. Academic Press, New York (1964)

Chapter 2
Some Arithmetic and Analysis in $\mathbb{Q}_p$; Derivatives in Ultrametric Analysis

Abstract In this chapter, we discuss some arithmetic and analysis in the p-adic field. We also introduce the concepts of differentiability and derivatives in ultrametric analysis and briefly indicate how ultrametric calculus is different from our usual calculus.

Keywords Valuation ring · Residue class field · Canonical expansion · Differentiability · Derivatives

We need the following results in the sequel.

Theorem 2.1 *Let $| \cdot |$ be an ultrametric valuation of K. Then the set $V \subseteq K$ of all elements a such that $|a| \leq 1$ is a ring with identity. The set $P \subseteq V$ of all elements a such that $|a| < 1$ is the unique maximal ideal of V and P is also a prime ideal.*

Proof If $a, b \in V$, then $|ab| = |a||b| \leq 1$ and so $ab \in V$. Again,

$$|a - b| \leq \max(|a|, |b|) \leq 1,$$

so that $a - b \in V$. Thus V is a ring. Also $1 \in V$. Now, if $a, b \in P$,

$$|a + b| \leq \max(|a|, |b|) < 1,$$

and so $a + b \in P$. If $a \in P, b \in V$, then

$$|ba| = |b||a| < 1,$$

so $ba \in P$. Thus P is an ideal of V. Further, if $a \in V$ and if $a \notin P$, then $|a| = 1$. Now,

$$1 = |1| = |a.a^{-1}| = |a||a^{-1}|,$$

so that $|a^{-1}| = 1$. It now follows that P is the unique maximal ideal of V. Since $1 \in V$, P is also a prime ideal, completing the proof. $\qquad\square$

© Springer India 2015

P.N. Natarajan, *An Introduction to Ultrametric Summability Theory*,
Forum for Interdisciplinary Mathematics 2, DOI 10.1007/978-81-322-2559-1_2

The ring V is called the "valuation ring" associated with the ultrametric valuation $| \cdot |$. The field V/P is called the associated "residue class field".

Theorem 2.2 *If $| \cdot |$ is an ultrametric valuation of K and $\hat{K}$ is the completion of K, then $|K| = |\hat{K}|$, where $|K|$ is the image of K in $\mathbb{R}$ under the valuation $| \cdot |$ and $|\hat{K}|$ is the image of $\hat{K}$ in $\mathbb{R}$ under the extended valuation (for the notion of extended valuation, see [1]), which we denote by $| \cdot |$ again.*

Proof Let $\alpha \in \hat{K}$. If $\alpha = 0$, $|\alpha| = 0$. So let $\alpha \neq 0$. Since K is dense in $\hat{K}$, there exists a sequence $\{a_n\}$ in K such that $\lim_{n \to \infty} a_n = \alpha$. Now, since $| \cdot |$ is an ultrametric valuation,

$$|a_n| = |\alpha + (a_n - \alpha)| = \max(|\alpha|, |a_n - \alpha|) = |\alpha|,$$

for sufficiently large n, since $|\alpha| \neq 0$, $|a_n - \alpha|$ can be made arbitrarily small for sufficiently large n. Thus $|\hat{K}| \subseteq |K|$. The reverse inclusion is trivial. This proves the theorem.   □

Theorem 2.3 *Any $\alpha \in \mathbb{Q}_p$ can be written as*

$$\alpha = \sum_{j=n}^{\infty} a_j p^j, \tag{2.1}$$

where $a_j \in \mathbb{Z}$, $j = n, n+1, \ldots$ and n is such that $|\alpha|_p = |p|_p^n$.

Proof Let $\alpha \in Q_p$, $\alpha \neq 0$. In view of Theorem 2.2,

$$|\mathbb{Q}_p|_p = |\mathbb{Q}|_p = \{|p|_p^n, n = 0, \pm 1, \pm 2, \ldots\}$$

so that

$$|\alpha|_p = |p|_p^n \text{ for some } n \in \mathbb{Z}. \tag{2.2}$$

Let $\beta = \frac{\alpha}{p^n}$ so that $|\beta|_p = 1$. Let V, P respectively denote the valuation ring of $| \cdot |_p$ on $\mathbb{Q}$ and the unique maximal ideal of V; let $\hat{V}, \hat{P}$ respectively denote the valuation ring of $| \cdot |_p$ on $\mathbb{Q}_p$ and the unique maximal ideal of $\hat{V}$. Now, $\beta \in \hat{V}$ and so $\beta = \lim_{k \to \infty} c_k, c_k \in \mathbb{Q}, k = 0, 1, 2, \ldots$. There exists a positive integer N (depending on n) such that $|\beta - c_k|_p < 1, k \geq N$. In particular, $|\beta - c_N|_p < 1$. Consequently,

$$|c_N|_p = |\beta + (c_N - \beta)|_p = \max(|\beta|_p, |c_N - \beta|_p) = 1,$$

since $| \cdot |_p$ is an ultrametric valuation. Since $c_N \in \mathbb{Q}$ and $|c_N|_p = 1, c_N \in V$. Let us write $c_N = b_n$. $|\beta - b_n|_p = |\beta - c_N|_p < 1$ and so $\beta - b_n \in \hat{P}$. Thus $\beta + \hat{P} = b_n + \hat{P}$. Since $|b_n|_p = 1, b_n \in \mathbb{Q}, b_n = \frac{e_n}{d_n}, e_n, d_n \in \mathbb{Z}, e_n, d_n$ are prime to p. Thus there exist integers x, y such that

$$xd_n + yp = 1,$$

$$i.e., xd_n \equiv 1 (mod\ p).$$

Then,

$$b_n - e_n x = \frac{e_n}{d_n} - e_n x$$

$$= e_n \left(\frac{1}{d_n} - x \right)$$

$$= \frac{e_n(1 - d_n x)}{d_n}$$

$$\equiv 0 (mod\ p)$$

so that $|b_n - e_n x|_p < 1$. Thus $b_n - e_n x \in P$ and so $b_n - e_n x \in \hat{P}$. Let $a_n = e_n x$. Then $a_n \in \mathbb{Z}$ and $|b_n - a_n|_p < 1$, i.e., $b_n - a_n \in \hat{P}$ and so $b_n + \hat{P} = a_n + \hat{P}$. Already $\beta + \hat{P} = b_n + \hat{P}$ so that $\beta + \hat{P} = a_n + \hat{P}$. Thus $\beta - a_n \in \hat{P}$ and so $|\beta - a_n|_p < 1$. We note that $|a_n|_p = |(a_n - \beta) + \beta|_p = \max(|a_n - \beta|_p, |\beta|_p) = 1$. We now have

$$\alpha = \beta p^n = a_n p^n + (\beta - a_n) p^n$$
$$= a_n p^n + \gamma_1,$$

where $\gamma_1 = (\beta - a_n) p^n$.

$$|\gamma_1|_p = |(\beta - a_n) p^n|_p = |\beta - a_n|_p |p|_p^n < |p|_p^n.$$

So

$$|\gamma_1|_p = |p|_p^m, \quad \text{where } m > n, \tag{2.3}$$

which is similar to (2.2). Treating γ_1 like α and continuing the process, after k steps, we get

$$\alpha = a_n p^n + a_{n+1} p^{n+1} + \cdots + a_{n+k-1} p^{n+k-1} + \gamma_k,$$

where $a_i \in \mathbb{Z}, i = n, n+1, \ldots, n+k-1, |a_i|_p = 1$ or 0 and

$$|\gamma_k|_p < |p|_p^{n+k} \to 0, k \to \infty,$$

since $|p|_p < 1$. This completes the proof of the theorem.   □

The integer coefficients of (2.1) are only unique modulo p. So if we agree to choose the a_i's such that $0 \le a_i \le p - 1$, then (2.1) is called the "canonical representation or expansion" of α.

We shall illustrate the above with an example. We shall find the canonical expansion of $\frac{3}{8}$ in $\mathbb{Q}_5$. We shall follow the notation used in Theorem 2.3. Since $\left| \frac{3}{8} \right|_5 = |5|_5^0 = 1$, we see that $n = 0$. A solution of

$$8x \equiv 1 (mod\ 5)$$

is $x = 2$. Since $3 \cdot 2 \equiv 1 (mod\ 5)$, $a_0 = 1$. Now,

$$\gamma_1 = \left(\frac{3}{8} - 1\right) 5^0 = -\frac{5}{8}.$$

We repeat the above procedure for γ_1. $|\gamma_1|_5 = \left|-\frac{5}{8}\right|_5 = |5|_5^1$. Now, $-\frac{5}{8} \cdot \frac{1}{5} = -\frac{1}{8}$.
Again, a solution of

$$8x \equiv 1 (mod\ 5)$$

is $x = 2 \cdot (-1) \cdot 2 \equiv 3 (mod\ 5)$ so that $a_1 = 3$. Next,

$$\gamma_2 = \left(-\frac{1}{8} - 3\right) 5 = \left(-\frac{25}{8}\right) 5$$

and so $|\gamma_2|_5 = |5|_5^3$, which implies that $a_2 = 0$. Since $\frac{\gamma_2}{5^3} = -\frac{1}{8}$, proceeding as
above, we see that $a_3 = 3$. Continuing in this manner, we see that

$$a_4 = a_6 = a_8 = \cdots = 0 \ \text{ and } \ a_5 = a_7 = a_9 = \cdots = 3.$$

We now follow the notation as under:
    The expansion of $\alpha \in \mathbb{Q}_p$, say,

$$\alpha = \frac{a_{-\gamma}}{p^\gamma} + \frac{a_{-\gamma+1}}{p^{\gamma-1}} + \cdots + a_0 + a_1 p + a_2 p^2 + \cdots$$

is, for convenience, written as

$$\alpha = a_{-\gamma} a_{-\gamma+1} \ldots a_0, a_1 a_2 \ldots (p). \qquad (2.4)$$

We now write the canonical expansion of $\frac{3}{8}$ in $\mathbb{Q}_5$ as

$$\frac{3}{8} = 1, 30\ 30\ 30 \ldots (5),$$

or, in a shorter form as

$$\frac{3}{8} = 1, \overline{30} \ldots (5),$$

where the bar above denotes periodic repetition.
    If $\alpha \in \mathbb{Q}_p$ has an expansion of the form

$$\alpha = a_0, a_1 a_2 \ldots (p),$$

then α is called a p-adic integer. Note that α is a p-adic integer if and only if $|\alpha|_p = |p|_p^n$ with $n \geq 0$, i.e., if and only if $\alpha \in \hat{V}$, the valuation ring of $|\cdot|_p$ on $\mathbb{Q}_p$.

We shall now illustrate the arithmetic operations in $\mathbb{Q}_p$, using the notation introduced in (2.4).

Addition

(1) In $\mathbb{Q}_7$, add the following

$$
\begin{array}{r}
1\;1\;1\;\;\;1\;1\;\;\;1 \\
4\;5\;2,\;1\;3\;7\;6\;1\;2 \\
+\quad 3\;7,\;5\;2\;1\;3\;1\;5\;2 \\
\hline
4\;1\;3,\;0\;6\;1\;3\;3\;0\;3
\end{array}
$$

(2) In $\mathbb{Q}_5$,

$$
\begin{array}{r}
1,\;3\;0\;3\;0\;3\;0\ldots \quad (=\tfrac{3}{8}) \\
+\;1\;0,\;0\;0\;0\;0\;0\;0\ldots \quad (=\tfrac{1}{5}) \\
\hline
1\;1,\;3\;0\;3\;0\;3\;0\ldots
\end{array}
$$

Subtraction

(1) In $\mathbb{Q}_7$,

$$
\begin{array}{r}
5\;6,\;3\;5\;2\;4 \\
-\quad 1,\;2\;4\;0\;3 \\
\hline
5\;5,\;1\;1\;2\;1
\end{array}
$$

(2) In $\mathbb{Q}_5$,

$$
\begin{array}{r}
7\;5\;3\;\;\;5\;6 \\
2\;2\;1,\;4\;3\;0\;2\;1 \\
-\;1\;3\;4,\;2\;3\;1\;4\;2\;2 \\
\hline
1\;4\;1,\;1\;0\;4\;2\;3\;2\;4\;4\ldots
\end{array}
$$

Multiplication

In $\mathbb{Q}_7$,

$$
\begin{array}{r}
1\;2,\;3\;1\;4 \\
\times\quad 1,\;2\;0\;3 \\
\hline
1\;2,\;3\;1\;4 \\
2,\;4\;6\;2\;1\;1 \\
3\;6\;2\;4\;5\;1 \\
\hline
1\;4,\;0\;4\;6\;4\;5\;5\;1
\end{array}
$$

Division

In $\mathbb{Q}_5$, divide 32,13 by 43,12

```
4 3, 1 2 )  3  2,  1  3 ( 2, 0̅ 2̅ 4̅ 4̅ 2̅ 0̅
              3 2, 3 4
              ─────────
                3 3 4 4 ...
                3 2 3 4
                ─────────
                  1 1 0  4  4 ...
                  1 0 2  4  1
                  ───────────
                    1 3 4 2 4 4 ...
                    1 0 2 4 1
                    ─────────────
                      3 2 3 2 4 4 ...
                      3 2 3 4
                      ─────────────
                        3 3 4 4  ...
```

In $\mathbb{R}$, we write a given number as a decimal expansion. Its analogue in $\mathbb{Q}_p$ is the canonical expansion. We recall that in $\mathbb{R}$, a number is rational if and only if its decimal expansion is periodic. We have an analogue in $\mathbb{Q}_p$. We state the result (for proof, one can refer to [1]).

Theorem 2.4 *An element $\alpha \in \mathbb{Q}_p$ is rational if and only if its canonical expansion*

$$\alpha = \sum_{j=n}^{\infty} a_j p^j, \quad 0 \le a_j \le p - 1,$$

when n is such that $|\alpha|_p = |p|_p^n$, is periodic.

Example 2.1 Find the rational number represented by the canonical expansion $1, \overline{30}$ in $\mathbb{Q}_5$.

$$
\begin{aligned}
1, \overline{30} &= 1 + 3.5^1 + 0.5^2 + 3.5^3 + 0.5^4 + \cdots \\
&= 1 + 3[5^1 + 5^3 + \cdots] \\
&= 1 + 3 \cdot \frac{5}{1 - 5^2}, \quad \text{since } |5^2|_5 = |5|_5^2 < 1 \\
&= 1 - \frac{15}{24} \\
&= 1 - \frac{5}{8} \\
&= \frac{3}{8}.
\end{aligned}
$$

Exercise 2.1 Find the canonical expansion of

(i) $\frac{1}{5}$ in $\mathbb{Q}_3$;   (ii) $\frac{1}{3}$ in $\mathbb{Q}_2$;   (iii) $-\frac{5}{7}$ in $\mathbb{Q}_5$.

Exercise 2.2 In $\mathbb{Q}_5$, find

(i) $\begin{array}{l} 1\ 2\ 3,\ 4\ 1\ 2 \\ +\ 4\ 2\ 1,\ 0\ 3\ 2 \end{array}$;

(ii) $\begin{array}{l} 1\ 2\ 4,\ 1\ 3\ 1 \\ -\ 3\ 2\ 1,\ 2\ 2\ 1 \end{array}$;

(iii) $\begin{array}{l} (3\ 4,\ 1\ 2\ 1) \\ \times\ (0,\ 2\ 1\ 0\ 3) \end{array}$;

(iv) $(1\ 3\ 1,\ 2) \div (2,\ 4\ 2)$

Exercise 2.3 In $\mathbb{Q}_3$, find the rational number whose canonical expansion is $2, \overline{0121}$.

As in the classical set up, in $\mathbb{Q}_p$ too, we have the "exponential" and "logarithmic" series respectively defined by

$$E(x) = \sum_{n=0}^{\infty} \frac{x^n}{n!} \quad \text{and} \quad L(1+x) = \sum_{n=1}^{\infty} (-1)^{n-1} \frac{x^n}{n},$$

which converge for all $x \in \mathbb{Q}_p$ with $|x|_p < 1$. These series have properties which are very similar to their classical counterparts, say, for instance,

$E(x + y) = E(x)E(y); L((1 + x)(1 + y)) = L(1 + x) + L(1 + y);$
$L(E(x)) = x; E(L(1 + x)) = 1 + x.$

In $\mathbb{Q}_p$, we have Binomial series too (for details, refer to [1]).

Though not relevant to the present monograph, it is worth noting that the concept of derivative and its properties have been studied in ultrametric analysis (see [2]). With regard to derivatives, we need the following definition.

Definition 2.1 If U is any subset of an ultrametric field K without isolated points and $f : U \to K$, we say that f is differentiable at $x \in U$ if

$$\lim_{y \to 0} \frac{f(x + y) - f(x)}{y} \quad \text{exists.} \tag{2.5}$$

Whenever the limit (2.5) exists, it is called the derivative of f at x, denoted by $f'(x)$.

It is immediate from the above definition that the characteristic function (K-valued) χ_U of any clopen set (i.e., any set which is both open and closed) is differentiable everywhere with $\chi'_U = 0$ everywhere. This shows that there are nonconstant functions whose derivatives are 0 everywhere contrary to the classical situation. There exist (1-1) functions too whose derivatives are 0 everywhere. If the characteristic of K is 2 (We recall that any valuation of such a field is ultrametric, in view of Exercise 1.1) and $f : K \to K$ is defined by $f(x) = x^2$, $x \in K$, then f is (1-1) (since $f(x) = f(y) \Rightarrow x^2 = y^2 \Rightarrow x = y$, using the fact that the characteristic of K is 2) and

$$f'(a) = \lim_{x \to a} \frac{x^2 - a^2}{x - a} = \lim_{x \to a} (x + a) = 2a = 0,$$

at any $a \in K$. We can also give examples of functions which are continuous everywhere but not differentiable anywhere. For instance, let U denote the closed unit disc in $\mathbb{Q}_p$. Let $f : U \to \mathbb{Q}_p$ defined by $f\left(\sum_n a_n p^n\right) = \sum_n a_n p^{2n}, 0 \le a_n \le p - 1$.

Then f is continuous everywhere but not differentiable anywhere (see [3]). In classical analysis, functions which have antiderivatives do not have jump discontinuities and they are pointwise limits of continuous functions. However, both these conditions are not sufficient for the functions to have an antiderivative. Unlike the classical case in which sufficient conditions are not known, the situation in the ultrametric case is simpler: If U is a subset of K without isolated points, then $f : U \to K$ has an antiderivative if and only if f is the pointwise limit of continuous functions ([4], p. 283).

References

1. Bachman, G.: Introduction to p-adic Numbers and Valuation Theory. Academic Press, New York (1964)
2. Schikhof, W.H.: Ultrametric Calculus. Cambridge University Press, Cambridge (1984)
3. Narici, L., Beckenstein, E.: Strange terrain—non-archimedean spaces. Amer. Math. Monthly **88**, 667–676 (1981)
4. Van Rooij, A.C.M.: Non-archimedean Functional Analysis. Marcel Dekker, New York (1978)

Chapter 3
Ultrametric Functional Analysis

Abstract In this chapter, we introduce ultrametric Banach spaces and mention that many results of the classical Banach space theory carry over in the ultrametric set up too. However, the Hahn–Banach theorem fails to hold. To salvage the Hahn–Banach theorem, the concept of a "spherically complete field" is introduced and Ingleton's version of the Hahn–Banach theorem is proved. The classical "convexity" does not work in the ultrametric set up and it is replaced by the notion of "K-convexity", which is briefly discussed at the end of the chapter.

Keywords Ultrametric Banach space · Spherically complete filed · Hahn–Banach theorem · K-convexity

The definition of an ultrametric Banach space over an ultrametric field K is the same as for the classical case except that the norm satisfies the strong triangle inequality, i.e., $||x + y|| \leq \max(||x||, ||y||)$. We also require the condition that the valuation of K is non-trivial since with this assumption, a linear map between ultrametric Banach spaces is continuous if and only if it is bounded (note that this result may fail if the valuation of K is trivial). Even after excluding the trivial valuation of K, there may not exist unit vectors in X; a vector x cannot be divided by $||x||$, since $||x||$ is a real number and not a scalar. Because of these deficiencies, even one-dimensional ultrametric Banach spaces need not be isometrically isomorphic, though they are linearly homeomorphic.

Consider the following formulae for the norm of a bounded linear map $A : X \to Y$ where X, Y are ultrametric Banach spaces.

$$||A|| = \sup \left\{ \frac{||A(x)||}{||x||} : 0 < ||x|| \leq 1 \right\}; \tag{3.1}$$

$$||A|| = \sup \{||A(x)|| : ||x|| = 1\}; \tag{3.2}$$

$$||A|| = \sup \{||A(x)|| : 0 \leq ||x|| \leq 1\}. \tag{3.3}$$

P.N. Natarajan, *An Introduction to Ultrametric Summability Theory*,
Forum for Interdisciplinary Mathematics 2, DOI 10.1007/978-81-322-2559-1_3

We recall that in the classical case, all the above three formulae for norm hold. In the ultrametric case, if K is non-trivially valued, (3.1) holds but (3.2) and (3.3) may not hold ([1], p. 75).

Many results about classical and ultrametric Banach spaces have exactly same formal statements but the proofs are entirely different. For instance, consider the classical theorem that locally compact Banach spaces must be finite-dimensional. The proof of this theorem involves use of unit vectors. In the ultrametric case, we cannot use this technique. We consider various cases and vectors whose norms are close to 1 and finally conclude that locally compact ultrametric Banach spaces must be finite-dimensional and the underlying field K must be locally compact ([1], p. 70).

Some results of the classical Banach space theory which hold in the ultrametric setting too are the closed graph theorem, Banach-Steinhaus theorem and open mapping theorem. However, the classical Hahn–Banach theorem does not carry over to the ultrametric setting—this makes the situation more interesting. In the case of ultrametric Banach spaces, it may not be possible to extend a given continuous linear functional from a subspace to the entire space. The fault is not with the linear space X but with the underlying field K. With a view to retain the Hahn–Banach theorem in the ultrametric set up too, a new notion of "spherical completeness" is introduced. We recall that completeness means that every nested sequence of closed balls, whose diameters tend to 0, have non-empty intersection. Spherical completeness makes the same demand but drops the requirement that the diameters tend to 0, i.e., spherical completeness means that every nested sequence of closed balls have non-empty intersection. It is clear that spherical completeness is stronger than completeness. However, the converse is not true, a counterexample may not be easy! For a counterexample, see [1], Example 4, pp. 81–83. Ingleton proved the following result.

Theorem 3.1 ([1], p.78) (The ultrametric Hahn–Banach theorem) *Let X be a normed linear space and Y be an ultrametric normed linear space over K (K may even be trivially valued). A continuous linear mapping A defined on a subspace M of X into Y may be extended to a continuous linear mapping A^* of X into Y with the same norm, as defined by (3.1), if and only if Y is spherically complete.*

We now prove the above theorem to exhibit the crucial role played by "Spherical Completeness" in proving the ultrametric version of the Hahn–Banach theorem.

Proof of Theorem 3.1 Let A be a bounded linear transformation from M into Y, where $M \subseteq X$ and let Y be spherically complete. We will prove that A can be extended to the whole space X in a norm-preserving fashion. Let now $x_0 \in M'$ (the complement of M). Consider the subspace $[M, x_0]$ of X spanned by M and x_0. Let $C_{\epsilon_x}(Ax) = \{y \in Y / ||y - Ax|| \leq \epsilon_x\}$, where $\epsilon_x = ||A||_M ||x - x_0||$. For convenience, we shall use $||A||$ for $||A||_M$ henceforth. We claim that the set of spheres $\{C_{\epsilon_x}(Ax)/x \in M\}$ is a nest. Now,

$$||Ax_1 - Ax_2|| = ||A(x_1 - x_2)|| \le ||A||||x_1 - x_2||$$
$$\le ||A|| \max\{||x_1 - x_0||, ||x_2 - x_0||\},$$

i.e., $||Ax_1 - Ax_2|| \le \max\{\epsilon_{x_1}, \epsilon_{x_2}\}$.

It now follows that $Ax_1 \in C_{\epsilon_2}(Ax_2)$ or $Ax_2 \in C_{\epsilon_1}(Ax_1)$. Consequently, the spheres $\{C_{\epsilon_x}(Ax)/x \in M\}$ form a nest. Since Y is spherically complete, there exists $z_0 \in \bigcap_{x \in M} C_{\epsilon_x}(Ax)$. Define an extension $\bar{A}$ of A to $[M, x_0]$ by

$$\bar{A}(x + \lambda x_0) = Ax + \lambda z_0.$$

It is clear that $\bar{A}$ is a linear extension of A. Now,

$$||\bar{A}(x + \lambda x_0)|| = ||Ax + \lambda z_0||$$
$$= |\lambda|||A(\lambda^{-1}x) + z_0||$$
$$= |\lambda|||A(-\lambda^{-1}x) - z_0||$$
$$\le |\lambda|\epsilon_{-\lambda^{-1}x}$$
$$= |\lambda|||A||||-\lambda^{-1}x - x_0||$$
$$= ||A||||x + \lambda x_0||.$$

Thus $\bar{A}$ is bounded with $||\bar{A}|| = ||A||$. At this stage, an application of Zorn's lemma proves that A can be extended to the entire space X in a norm-preserving fashion.

To prove the converse, we shall suppose that Y is not spherically complete and arrive at a contradiction, i.e., exhibit a bounded linear transformation defined on a subspace of a normed linear space X, which cannot be extended to X in a norm-preserving fashion. Since we have assumed that Y is not spherically complete, there exists in Y a nonempty nest of spheres $\{C_{\epsilon_\alpha}(y_\alpha)/\alpha \in \Lambda\}$ with empty intersection. Let $y \in Y$. Then $y \notin C_{\epsilon_\beta}(y_\beta)$ for some $\beta \in \Lambda$. Define the function φ by

$$\varphi(y) = ||y - y_\beta||.$$

We now claim that φ is well defined. First we note that for any $z \in C_{\epsilon_\beta}(y_\beta)$, $||z - y_\beta|| \le \epsilon_\beta$. Since $y \notin C_{\epsilon_\beta}(y_\beta)$, $||y - y_\beta|| > \epsilon_\beta$ so that $||y - y_\beta|| > ||z - y_\beta||$. Thus

$$||y - z|| = ||(y - y_\beta) - (z - y_\beta)|| = ||y - y_\beta||,$$

using Theorem 1.1, so that

$$\varphi(y) = ||y - z||.$$

Suppose $y \notin C_{\epsilon_1}(y_1)$ and $y \notin C_{\epsilon_2}(y_2)$. Since the spheres form a nest, we can suppose that $C_{\epsilon_1}(y_1) \subseteq C_{\epsilon_2}(y_2)$ and so $\epsilon_1 \le \epsilon_2$. Now, $\varphi(y) = ||y - y_2|| > \epsilon_2$ and $||y_1 - y_2|| \le \epsilon_2$ so that

$$||y - y_1|| = ||(y - y_2) + (y_2 - y_1)|| = ||y - y_2||,$$

in view of Theorem 1.1. Thus

$$\varphi(y) = ||y - y_2|| = ||y - y_1||,$$

proving that φ is well defined.

Let now $y \in C_{\epsilon_\alpha}(y_\alpha)$. We have already noted that there exists a sphere $C_{\epsilon_\beta}(y_\beta)$ such that $y \notin C_{\epsilon_\beta}(y_\beta)$. Since the spheres form a nest and $y \in C_{\epsilon_\alpha}(y_\alpha)$, it follows that $C_{\epsilon_\beta}(y_\beta) \subseteq C_{\epsilon_\alpha}(y_\alpha)$. Consequently,

$$\varphi(y) = ||y - y_\beta|| \leq \max\{||y - y_\alpha||, ||y_\alpha - y_\beta||\}$$
$$\leq \epsilon_\alpha.$$

Thus, for $y \in C_{\epsilon_\alpha}(y_\alpha)$, $\varphi(y) \leq \epsilon_\alpha$. Let X be the direct sum of Y and F. For $(y, \lambda) \in X$, define

$$||(y, \lambda)|| = \begin{cases} |\lambda|\varphi(\lambda^{-1}y), & \text{if } \lambda \neq 0; \\ ||y||, & \text{if } \lambda = 0. \end{cases}$$

We prove that $|| \cdot ||$ defines a non-archimedean norm on X. Note that $\varphi(y) > 0$ implies that $||(y, \lambda)|| = 0$ if and only if $(y, \lambda) = (0, 0)$. It is also clear that

$$||\mu(y, \lambda)|| = ||(\mu y, \mu \lambda)|| = |\mu|||(y, \lambda)||.$$

It remains to prove that $|| \cdot ||$ satisfies the ultrametric inequality. We have to consider several cases. We will consider one of the typical cases, leaving the remaining ones as exercise. Let $(y_1, \lambda_1), (y_2, \lambda_2)$ be such that $\lambda_1, \lambda_2, \lambda_1 + \lambda_2 \neq 0$. We can find a sphere $C_{\epsilon_0}(y_0)$ from the nest such that $\lambda_1^{-1}y_1, \lambda_2^{-1}y_2, (\lambda_1 + \lambda_2)^{-1}(y_1 + y_2) \notin C_{\epsilon_0}(y_0)$. Now,

$$||(y_1, \lambda_1)|| = ||y_1 - \lambda_1 y_0||, ||(y_2, \lambda_2)|| = ||y_2 - \lambda_2 y_0||$$

and

$$||(y_1 + y_2, \lambda_1 + \lambda_2)|| = ||y_1 + y_2 - (\lambda_1 + \lambda_2)y_0||$$
$$\leq \max\{||(y_1, \lambda_1)||, ||(y_2, \lambda_2)||\},$$

completing the proof in this case.

Note that X contains an isometrically isomorphic image of Y, viz., $\hat{Y} = \{(y, 0)/y \in Y\}$. Hereafter, we will replace Y by $\hat{Y}$. Consider the linear transformation 1, defined by

$$1: \begin{array}{c} \hat{Y} \rightarrow \hat{Y} \\ (y, 0) \rightarrow (y, 0). \end{array}$$

Suppose 1 has a norm-preserving extension $\hat{1}$ to the entire space X. Note that $\|\hat{1}\| = 1$. Let $z \in Y$ be such that

$$\hat{1}(0, -1) = (z, 0).$$

$$\hat{1}(y, 1) = \hat{1}(y, 0) + \hat{1}(0, 1) = (y, 0) - (z, 0) = (y - z, 0).$$

Now,

$$\|y - z\| = \|\hat{1}(y, 1)\| \le \|(y, 1)\|, \ \text{using} \ \|\hat{1}\| = 1$$
$$= \varphi(y).$$

If we consider the vector y_α from each of the spheres $C_{\epsilon_\alpha}(y_\alpha)$ of the nest, we get $\|y_\alpha - z\| \le \varphi(y_\alpha) \le \epsilon_\alpha$. Thus $z \in \bigcap_{\alpha \in \Lambda} C_{\epsilon_\alpha}(y_\alpha)$, a contradiction of our assumption that $\bigcap_{\alpha \in \Lambda} C_{\epsilon_\alpha}(y_\alpha) = \phi$, completing the proof of the theorem. □

There are equivalent ways of describing spherical completeness ([1], Chap. 2). The idea that spherical completeness should be substituted in the ultrametric set up whenever completeness appears in the classical case is not, however, true. In fact, spherical completeness plays a very little role in the case of ultrametric Banach algebras and ultrametric measure theory. In fact, in some ways, it is disadvantageous: For instance, if K is spherically complete, then no infinite-dimensional ultrametric Banach space over K is reflexive ([2], Chap. 4). On the other hand, if K is not spherically complete, then c_0 and ℓ_∞ are reflexive.

Since ultrametric fields are totally disconnected, they are not totally ordered. The lack of ordering on K makes it difficult to find an analogue for classical "convexity". Classical convexity is replaced, in the ultrametric setting, by a notion called "K-convexity" (see [3]), which is defined as follows.

Definition 3.1 A set S of vectors is said to be "absolutely K-convex" if $ax + by \in S$ whenever $|a|, |b| \le 1$ and $x, y \in S$; translates $w + S$ of such sets S are called "K-convex". A topological linear space X (which is defined as in the classical case) is said to be "locally K-convex" if its topology has a base of K-convex sets at 0.

Using the above notion of K-convexity, a weaker form of compactness, called "c-compactness" could be defined. c-compactness demands that filterbases composed of K-convex sets possess adherence points rather than the usual requirement of compactness that all filterbases possess adherence points (see [4]). It is known that in the case of ultrametric fields, spherical completeness is equivalent to c-compactness. For a detailed study of K-convexity and c-compactness, one can refer to [3–5].

A seminorm p is ultrametric if it is a seminorm in the usual sense and if, in addition, satisfies $p(x + y) \le \max(p(x), p(y))$. The connection between K-convex sets and ultrametric seminorms is very similar to the classical case. The notion of K-convexity leads to defining F-barrels etc. These notions are used in spaces of continuous functions and analogues of known classical theorems are obtained.

Besides its requirements about complex conjugates, the inner product (,) on a classical Hilbert space satisfies the condition $(x, x) \geq 0$ for every x. Even assuming that an inner product on a linear space X over an ultrametric field K is scalar valued, the "≥ 0" cannot be carried over to the ultrametric set up. K is not totally ordered too. Because of these deficiencies, it seems that we do not have a meaningful analogue in the ultrametric setting of a classical Hilbert space. However, we can define the notion of "orthogonality" in ultrametric Banach spaces. This leads to the notion of an "orthogonal base", which has many properties as its classical counterpart.

For the theory of Banach algebras, spaces of continuous functions, Measure and integral, the reader can refer to [6] and the relevant references given in [6]. For some applications of ultrametric analysis to mathematical physics, one can refer to [7].

References

1. Narici, L., Beckenstein, E., Bachman, G.: Functional Analysis and Valuation Theory. Marcel Dekkar, New York (1971)
2. Van Rooij, A.C.M.: Non-archimedean Functional Analysis. Marcel Dekker, New York (1978)
3. Monna, A.F.: Ensembles convexes dans les espaces vectoriels sur un corps valué. Indag. Math. **61**, 528–539 (1958)
4. Springer, T.A.: Une notion de compacité dans la theéorie des espaces vectoriels topologiques. Indag. Math. **27**, 182–189 (1965)
5. van Tiel, J.: Espaces localement K-convexes (I–IV). Indag. Math. **27**, 249–289 (1965)
6. Narici, L., Beckenstein, E.: Strange terrain—non-archimedean spaces. Amer. Math. Monthly **88**, 667–676 (1981)
7. Vladimirov, V.S., Volovich, I.V., Zelenov, E.I.: p-adic analysis and mathematical physics. World Scientific, Singapore (1994)

Chapter 4
Ultrametric Summability Theory

Abstract Our survey of the literature on "ultrametric summability theory" starts with a paper of Andree and Petersen of 1956 (it was the earliest known paper on the topic) to the present. In the present chapter, Silverman–Toeplitz theorem is proved using the "sliding-hump method". Schur's theorem and Steinhaus theorem also find a mention. Core of a sequence and Knopp's core theorem is discussed. It is proved that certain Steinhaus-type theorems fail to hold.

Keywords Silverman–Toeplitz theorem · Schur's theorem · Steinhaus theorem · Steinhaus-type theorems · Knopp's core theorem

4.1 Classes of Matrix Transformations

We now present for the first time a brief survey, though not exhaustive, of the work done so far on ultrametric summability theory. Divergent series have been the motivating factor for the introduction of summability theory both in classical as well as ultrametric analysis.

Study of infinite matrix transformations in the classical case is quite an old one. Numerous authors have studied general matrix transformations or matrix transformations pertaining to some special classes for the past several decades. On the other hand, the study of matrix transformations in the ultrametric case is of a comparatively recent origin. In spite of the pioneering work of A.F. Monna relating to analysis and functional analysis over ultrametric fields dating back to 1940s, it was only in 1956 that Andree and Petersen [1] proved the analogue for the p-adic field $\mathbb{Q}_p$ of the Silverman–Toeplitz theorem on the regularity of an infinite matrix transformation. Roberts [2] proved the Silverman–Toeplitz theorem for a general ultrametric field, while Monna ([3], p. 127) obtained the same theorem using the Banach–Steinhaus theorem. Later to Monna, we have only the papers by Rangachari and Srinivasan [4], Srinivasan [5] and Somasundaram [6, 7] till Natarajan took up a detailed study of matrix transformations, special methods of summability and other aspects of summability theory in the ultrametric set up.

© Springer India 2015

P.N. Natarajan, *An Introduction to Ultrametric Summability Theory*,
Forum for Interdisciplinary Mathematics 2, DOI 10.1007/978-81-322-2559-1_4

In the sequel, K is a complete, non-trivially valued, ultrametric field unless otherwise stated. We shall suppose that the entries of sequences, series and infinite matrices are in K.

Definition 4.1 If $A = (a_{nk})$, $a_{nk} \in K$, $n, k = 0, 1, 2, \ldots$, is an infinite matrix, by the A-transform Ax of a sequence $x = \{x_k\}$, $x_k \in K$, $k = 0, 1, 2, \ldots$, we mean the sequence $\{(Ax)_n\}$, where

$$(Ax)_n = \sum_{k=0}^{\infty} a_{nk} x_k, \quad n = 0, 1, 2, \ldots,$$

it being assumed that the series on the right converge. If $\lim_{n \to \infty} (Ax)_n = \ell$, we say that the sequence $x = \{x_k\}$ is A-summable or summable A to ℓ.

Definition 4.2 Let X and Y be sequence spaces with elements whose entries are in K. The infinite matrix $A = (a_{nk})$, $a_{nk} \in K$, $n, k = 0, 1, 2, \ldots$ is said to transform X to Y if whenever the sequence $x = \{x_k\} \in X$, $(Ax)_n$ is defined, $n = 0, 1, 2, \ldots$ and the sequence $\{(Ax)_n\} \in Y$. In this case, we write $A \in (X, Y)$.

Definition 4.3 If $A \in (c, c)$ (where c is the ultrametric Banach space consisting of all convergent sequences in K with respect to the norm defined by $||x|| = \sup_{k \geq 0} |x_k|$, $x = \{x_k\} \in c$), A is said to be convergence preserving or conservative. If, in addition, $\lim_{n \to \infty} (Ax)_n = \lim_{k \to \infty} x_k$, A is called a regular matrix or a Toeplitz matrix. If A is regular, we write $A \in (c, c; P)$ (P standing for "preservation of limit").

Theorem 4.1 (Silverman–Toeplitz) ([3], p. 127) $A \in (c, c)$, *i.e.* A *is convergence preserving if and only if*

$$\sup_{n,k} |a_{nk}| < \infty; \tag{4.1}$$

$$\lim_{n \to \infty} a_{nk} = \delta_k \text{ exists, } k = 0, 1, 2, \ldots; \tag{4.2}$$

and

$$\lim_{n \to \infty} \sum_{k=0}^{\infty} a_{nk} = \delta \text{ exists.} \tag{4.3}$$

In such a case $\lim_{n \to \infty} (Ax)_n = s\delta + \sum_{k=0}^{\infty} (x_k - s)\delta_k$, $\lim_{k \to \infty} x_k = s$. Further, $A \in (c, c; P)$, i.e. A is regular if and only if (4.1), (4.2) and (4.3) hold with the additional requirements $\delta_k = 0$, $k = 0, 1, 2, \ldots$ and $\delta = 1$.

Remark 4.1 Monna [3] proved Theorem 4.1 using modern tools like the ultrametric version of Banach–Steinhas theorem. In this context, Natarajan [8] obtained Theorem 4.1, using the "sliding hump method". This proof is an instance in which traditional tools like 'signum function' available in $\mathbb{R}$ or $\mathbb{C}$ are also avoided.

We now present this proof.

Proof of Theorem 4.1 (see [8]) Let (4.1), (4.2) and (4.3) hold. In view of (4.1), there exists $H > 0$ such that

$$\sup_{n,k} |a_{nk}| \le H.$$

Using (4.2), $|\delta_k| \le H$, $k = 0, 1, 2, \ldots$. Let $\{x_k\}$ be such that $\lim_{k \to \infty} x_k = s$. Since $x_k - s \to 0$, $k \to \infty$ and $\{\delta_k\}$ is bounded, $\delta_k(x_k - s) \to 0$, $k \to \infty$ and so $\sum_{k=0}^{\infty} \delta_k(x_k - s)$ converges. Again, since $a_{nk} \to 0$, $k \to \infty$ using (4.3) and $\{x_k\}$ is bounded, $\sum_{k=0}^{\infty} a_{nk} x_k$ converges and so

$$(Ax)_n = \sum_{k=0}^{\infty} a_{nk} x_k, \quad n = 0, 1, 2, \ldots$$

is defined and

$$(Ax)_n = \sum_{k=0}^{\infty} (a_{nk} - \delta_k)(x_k - s) + \sum_{k=0}^{\infty} \delta_k(x_k - s) + s \sum_{k=0}^{\infty} a_{nk}.$$

Since $x_k - s \to 0$, $k \to \infty$, given $\epsilon > 0$, there exists a positive integer k_0 such that

$$|x_k - s| < \frac{\epsilon}{H}, \quad k > k_0.$$

Now,

$$\sum_{k=0}^{\infty} (a_{nk} - \delta_k)(x_k - s) = \sum_{k=0}^{k_0} (a_{nk} - \delta_k)(x_k - s) + \sum_{k=k_0+1}^{\infty} (a_{nk} - \delta_k)(x_k - s).$$

By (4.2), $a_{nk} - \delta_k \to 0$, $n \to \infty$, $k = 0, 1, 2, \ldots$ and so we can find a positive integer N such that

$$|a_{nk} - \delta_k| < \frac{\epsilon}{M}, \quad n > N, k = 0, 1, 2, \ldots, k_0,$$

$M = \sup_{k \ge 0} |x_k - s|$. Thus

$$\left| \sum_{k=0}^{k_0} (a_{nk} - \delta_k)(x_k - s) \right| < \frac{\epsilon}{M} \cdot M = \epsilon, \quad n > N.$$

Also

$$\left| \sum_{k=k_0+1}^{\infty} (a_{nk} - \delta_k)(x_k - s) \right| \leq \sup_{k>k_0} |a_{nk} - \delta_k| \, |x_k - s|$$

$$< H \cdot \frac{\epsilon}{H} = \epsilon, \quad n = 0, 1, 2, \ldots.$$

So

$$\left| \sum_{k=0}^{\infty} (a_{nk} - \delta_k)(x_k - s) \right| < \epsilon, \quad n > N,$$

from which it follows that

$$\lim_{n \to \infty} \sum_{k=0}^{\infty} (a_{nk} - \delta_k)(x_k - s) = 0.$$

Consequently,

$$\lim_{n \to \infty} (Ax)_n = \sum_{k=0}^{\infty} \delta_k (x_k - s) + s\delta,$$

using (4.3).

Conversely, the necessity of conditions (4.2) and (4.3) are clear by considering the sequences $\{0, 0, \ldots, 0, 1, 0, \ldots\}$, 1 occurring in the kth place, $k = 0, 1, 2, \ldots$ and $\{1, 1, 1, \ldots\}$. We will now prove the necessity of (4.1). Using (4.3), we first note that

$$\sup_{k \geq 0} |a_{n+1,k} - a_{nk}| < \infty, \quad n = 0, 1, 2, \ldots. \tag{4.4}$$

We now claim that

$$\sup_{n,k} |a_{n+1,k} - a_{nk}| < \infty. \tag{4.5}$$

If not, we can then choose a strictly increasing sequence $\{n(m)\}$ of positive integers such that

$$\sup_{k \geq 0} |a_{n(m)+1,k} - a_{n(m),k}| > m^2, \quad m = 1, 2, \ldots. \tag{4.6}$$

In particular,

$$\sup_{k \geq 0} |a_{n(1)+1,k} - a_{n(1),k}| > 1^2. \tag{4.7}$$

Since $a_{n(1)+1,k} - a_{n(1),k} \to 0$, $k \to \infty$, given $\epsilon > 0$ (we can suppose that $\epsilon < 1$), we can choose a positive integer $k(n(1))$ such that

$$\sup_{k \geq k(n(1))} |a_{n(1)+1,k} - a_{n(1),k}| < \epsilon. \tag{4.8}$$

From (4.7) and (4.8), we have

$$\sup_{0 \leq k < k(n(1))} |a_{n(1)+1,k} - a_{n(1),k}| > 1^2,$$

so that there exists an integer $k(1)$, $0 \leq k(1) < k(n(1))$ with

$$|a_{n(1)+1,k(1)} - a_{n(1),k(1)}| > 1^2. \tag{4.9}$$

From (4.6),

$$\sup_{k \geq 0} |a_{n(2)+1,k} - a_{n(2),k}| > 2^2. \tag{4.10}$$

Since, by (4.2),

$$\lim_{n \to \infty} (a_{n+1,k} - a_{nk}) = 0, \quad k = 0, 1, 2, \ldots,$$

we may suppose that

$$\sup_{0 \leq k < k(n(1))} |a_{n(2)+1,k} - a_{n(2),k}| < \epsilon. \tag{4.11}$$

Since $a_{n(2)+1,k} - a_{n(2),k} \to 0$, $k \to \infty$, we can choose a positive integer $k(n(2)) > k(n(1))$ such that

$$\sup_{k \geq k(n(2))} |a_{n(2)+1,k} - a_{n(2),k}| < \epsilon. \tag{4.12}$$

Using (4.10), (4.11) and (4.12), it follows that

$$\sup_{k(n(1)) \leq k < k(n(2))} |a_{n(2)+1,k} - a_{n(2),k}| > 2^2.$$

So there exists a positive integer $k(2)$ such that

$$|a_{n(2)+1,k(2)} - a_{n(2),k(2)}| > 2^2. \tag{4.13}$$

Proceeding in this fashion, we have strictly increasing sequences $\{n(m)\}$, $\{k(n(m))\}$, $\{k(m)\}$ of positive integers with $k(n(m-1)) \leq k(m) < k(n(m))$ such that

$$\begin{cases} \text{(i)} \quad \sup_{0 \leq k < k(n(m-1))} |a_{n(m)+1,k} - a_{n(m),k}| < \epsilon; \\ \text{(ii)} \quad \sup_{k(n(m)) \leq k < \infty} |a_{n(m)+1,k} - a_{n(m),k}| < \epsilon; \\ \text{and} \\ \text{(iii)} \quad |a_{n(m)+1,k(m)} - a_{n(m),k(m)}| > m^2, \quad m = 1, 2, \ldots. \end{cases} \tag{4.14}$$

Since K is non-trivially valued, for $m = 1, 2, \ldots$, we can choose a non-negative integer $\alpha(m)$ such that

$$\rho^{\alpha(m)+1} < \frac{1}{m} \leq \rho^{\alpha(m)}, \tag{4.15}$$

where $\rho = |\pi| < 1$, $\pi \in K$. Define the sequence $x = \{x_k\}$, where

$$x_k = \pi^{\alpha(m)}, \quad \text{if } k = k(m), m = 1, 2, \ldots;$$
$$= 0, \text{ otherwise.}$$

It is clear that $\{x_k\}$ converges (in fact to 0). Using (4.14)(iii) and (4.15),

$$m < |a_{n(m)+1,k(m)} - a_{n(m),k(m)}||x_{k(m)}|.$$

Again, using (4.14)(i)–(ii),

$$|a_{n(m)+1,k(m)} - a_{n(m),k(m)}| \leq \max\{|(Ax)_{n(m)+1} - (Ax)_{n(m)}|, \epsilon\}.$$

Thus

$$m < \max\{|(Ax)_{n(m)+1} - (Ax)_{n(m)}|, \epsilon\},$$

which, in turn, implies that

$$|(Ax)_{n(m)+1} - (Ax)_{n(m)}| > m, \quad m = 1, 2, \ldots.$$

So $\{(Ax)_n\}$ is not a Cauchy sequence and does not therefore converge, which is a contradiction. This proves that (4.5) holds. Let

$$\sup_{n,k} |a_{n+1,k} - a_{nk}| < M_1, \quad M_1 > 0.$$

Since $a_{1k} \to 0$, $k \to \infty$, $|a_{1k}| < M_2$, for some $M_2 > 0$, $k = 0, 1, 2, \ldots$. Let $M = \max(M_1, M_2)$. Then

$$|a_{2k}| \leq \max\{|a_{2k} - a_{1k}|, |a_{1k}|\} \leq M, \quad k = 0, 1, 2, \ldots;$$
$$|a_{3k}| \leq \max\{|a_{3k} - a_{2k}|, |a_{2k}|\} \leq M, \quad k = 0, 1, 2, \ldots;$$

Inductively,

$$\sup_{n,k} |a_{nk}| \leq M,$$

which shows that (4.1) holds. The latter part of the theorem relating to the regularity of the method is easily deduced thereafter. This completes the proof of the theorem.

□

Definition 4.4 A is called a Schur matrix if $A \in (\ell_\infty, c)$ (where ℓ_∞ denotes the ultrametric Banach space of all bounded sequences in K with respect to the norm defined by $||x|| = \sup_{k \geq 0} |x_k|$, $x = \{x_k\} \in \ell_\infty$).

Natarajan [9] proved the following result.

Theorem 4.2 (Schur) $A \in (\ell_\infty, c)$, i.e. A is a Schur matrix if and only if

$$\lim_{k \to \infty} a_{nk} = 0, \quad n = 0, 1, 2, \ldots; \tag{4.16}$$

and

$$\lim_{n \to \infty} \sup_{k \geq 0} |a_{n+1,k} - a_{nk}| = 0. \tag{4.17}$$

The following result is immediately deduced [9].

Theorem 4.3 (Steinhaus) *A matrix cannot be both a Toeplitz and a Schur matrix, or, equivalently, given any regular matrix A, there exists a bounded divergent sequence which is not summable A. Symbolically, we can write this statement as $(c, c; P) \cap (\ell_\infty, c) = \phi$.*

Somasundaram [6] obtained the Steinhaus theorem for a restricted class of regular matrices and observed that the theorem was not true in general. His restriction is in fact, meaningless and his observation is incorrect. Neither of the examples in Sect. 2 of [6] provide a regular matrix which sums all bounded sequences.

4.2 Steinhaus-Type Theorems

In the context of Steinhaus theorem, we introduce the following.

Definition 4.5 Whatever be K, i.e. $K = \mathbb{R}$ or $\mathbb{C}$ or a complete, non-trivially valued, ultrametric field, Λ_r is the space of all sequences $x = \{x_k\} \in \ell_\infty$ such that $|x_{k+r} - x_k| \to 0, k \to \infty, r \geq 1$ being a fixed integer.

It is easily proved that Λ_r is a closed subspace of ℓ_∞ with respect to the norm defined for elements in ℓ_∞. When $K = \mathbb{R}$ or $\mathbb{C}$, the following result, improving Steinhaus theorem, was proved in [10] (it is worthwhile to note that a constructive proof was given).

Theorem 4.4 $(c, c; P) \cap (\Lambda_r - \bigcup_{i=1}^{r-1} \Lambda_i, c) = \phi.$

Proof Let $A = (a_{nk})$ be a regular matrix. We can now choose two sequences of positive integers $\{n(m)\}$ and $\{k(m)\}$ such that if

$$m = 2p, \ n(m) > n(m-1), \ k(m) > k(m-1) + (2m-5)r,$$

then

$$\sum_{k=0}^{k(m-1)+(2m-5)r} |a_{n(m),k}| < \frac{1}{16},$$

$$\sum_{k=k(m-1)}^{\infty} |a_{n(m),k}| < \frac{1}{16};$$

and if

$$m = 2p+1, \ n(m) > n(m-1), \ k(m) > k(m-1) + (m-2)r,$$

then

$$\sum_{k=0}^{k(m-1)+(m-2)r} |a_{n(m),k}| < \frac{1}{16},$$

$$\sum_{k=k(m-1)+(m-2)r+1}^{k(m)} |a_{n(m),k}| > \frac{7}{8},$$

$$\sum_{k=k(m)+1}^{\infty} |a_{n(m),k}| < \frac{1}{16}.$$

Define the sequence $x = \{x_k\}$ as follows:

if $k(2p-1) < k \le k(2p)$, then

$$x_k = \begin{cases} \frac{2p-2}{2p-1}, & k = k(2p-1)+1, \\ 1, & k(2p-1)+1 < k \le k(2p-1)+r, \\ \frac{2p-3}{2p-1}, & k = k(2p-1)+r+1, \\ 1, & k = k(2p-1)+r+1 < k \le k(2p-1)+2r, \\ \vdots \\ 1, & k(2p-1)+(2p-4)r+1 < k \le k(2p-1)+(2p-3)r, \\ \frac{1}{2p-1}, & k = k(2p-1)+(2p-3)r+1, \\ \frac{2p-2}{2p-1}, & k(2p-1)+(2p-3)r+1 < k \le k(2p-1)+(2p-2)r, \\ 0, & k = k(2p-1)+(2p-2)r+1, \\ \vdots \\ \frac{1}{2p-1}, & k(2p-1)+(4p-6)r+1 < k \le k(2p-1)+(4p-5)r, \\ 0, & k(2p-1)+(4p-5)r < k \le k(2p), \end{cases}$$

and if $k(2p) < k \le k(2p+1)$, then

$$x_k = \begin{cases} \frac{1}{2p}, & k(2p) < k \le k(2p)+r, \\ \frac{2}{2p}, & k(2p)+r < k \le k(2p)+2r, \\ \vdots \\ \frac{2p-1}{2p}, & k(2p)+(2p-2)r < k \le k(2p)+(2p-1)r, \\ 1, & k(2p)+(2p-1)r < k \le k(2p+1). \end{cases}$$

We note that, if $k(2p-1) < k \le k(2p)$,

$$|x_{k+r} - x_k| < \frac{1}{2p-1},$$

while, if $k(2p) < k \le k(2p+1)$,

$$|x_{k+r} - x_k| < \frac{1}{2p}.$$

Thus $|x_{k+r} - x_k| \to 0, k \to \infty$, showing that $x = \{x_k\} \in \Lambda_r$.
However,

$$|x_{k+1} - x_k| = \frac{2p-2}{2p-1}, \quad \text{if } k = k(2p-1)+(2p-3)r, \ p = 1, 2, \ldots.$$

Hence, $|x_{k+1} - x_k| \not\to 0, k \to \infty$ and consequently $x \notin \Lambda_1$. In a similar manner, we can prove that $x \notin \Lambda_i, i = 2, 3, \ldots, (r-1)$. Thus $x \in \Lambda_r - \bigcup_{i=1}^{r-1} \Lambda_i$. Further,

$$\left. \begin{array}{l} \left| (Ax)_{n(2p)} \right| < \frac{1}{16} + \frac{1}{16} = \frac{1}{8}, \\ \left| (Ax)_{n(2p+1)} \right| > \frac{7}{8} - \frac{1}{16} - \frac{1}{16} = \frac{3}{4} \end{array} \right\} p = 1, 2, \ldots,$$

which shows that $\{(Ax)_n\} \notin c$, completing the proof of the theorem. $\square$

In this context, it is worthwhile to note that the analogue of Theorem 4.4 in the ultrametric case fails to hold, as the following example shows. Consider the matrix $A = (a_{nk}), a_{nk} \in Q_p$, the p-adic field for a prime $p, n, k = 0, 1, 2, \ldots$, where

$$\left. \begin{array}{l} a_{nk} = \frac{1}{r}, \ k = n, n+1, \ldots, n+r-1 \\ = 0, \text{ otherwise} \end{array} \right\}, n = 0, 1, 2, \ldots,$$

r being a fixed positive integer.

$$(Ax)_{n+1} - (Ax)_n = \frac{(x_{n+1} + x_{n+2} + \cdots + x_{n+r}) - (x_n + x_{n+1} + \cdots + x_{n+r-1})}{r}$$

$$= \frac{(x_{n+r} - x_n)}{r} \to 0, \; n \to \infty,$$

if $x = \{x_k\} \in \Lambda_r$. Thus A sums all sequences in Λ_r. It is clear that A is regular. Consequently, $(c, c; P) \cap (\Lambda_r, c) \neq \phi$, proving our claim.

We note that $(c, c; P) \cap (\Lambda_r, c) = \phi$ when $K = \mathbb{R}$ or $\mathbb{C}$. Since $(\ell_\infty, c) \subseteq (\Lambda_r, c)$, it follows that $(c, c; P) \cap (\ell_\infty, c) = \phi$, which is Steinhaus theorem. We call such results Steinhaus-type theorems. For more Steinhaus-type theorems, see [11–13].

Let us now see in detail the role played by the sequence spaces Λ_r. Whatever be K, it is well known that an infinite matrix which sums all sequences of 0's and 1's sums all bounded sequences (see [14, 15]). It is clear that any Cauchy sequence is in $\bigcap\limits_{r=1}^{\infty} \Lambda_r$ so that each Λ_r is a sequence space containing the space $\mathscr{C}$ of Cauchy sequences. It may be noted that $\mathscr{C} \subsetneqq \bigcap\limits_{r=1}^{\infty} \Lambda_r$ when $K = \mathbb{R}$ or $\mathbb{C}$ while $\mathscr{C} = \bigcap\limits_{r=1}^{\infty} \Lambda_r$ when K is a complete, non-trivially valued, ultrametric field. Although Λ_r do not form a tower between $\mathscr{C}$ and ℓ_∞, they can be deemed to reflect the measure of non-Cauchy nature of sequences contained in them. It is also easy to prove that $\Lambda_r \subseteq \Lambda_s$ if and only if s is a multiple of r and that $\Lambda_r \cap \Lambda_{r+1} = \Lambda_1$. It is worthwhile to observe the nature of location of sequences of 0's and 1's in these spaces Λ_r. In the first instance, we note that a sequence of 0's and 1's is in Λ_r if and only if it is periodic with period r eventually. Consequently, any sequence of 0's and 1's is in $\ell_\infty - \bigcup\limits_{r=1}^{\infty} \Lambda_r$ if and only if it is non-periodic. If $\mathscr{NP}$ denotes the set of all sequences of 0's and 1's in $\ell_\infty - \bigcup\limits_{r=1}^{\infty} \Lambda_r$, it is proved in [14] that any infinite matrix which sums all sequences of $\mathscr{NP}$ is a Schur matrix whatever be K. This result is an improvement of Steinhaus theorem. If K is further locally compact (this requirement is fulfilled when $K = \mathbb{R}$ or $\mathbb{C}$), the closed linear span of $\mathscr{NP}$ is ℓ_∞.

Following Sember and Freedman [15], we now make a further study of sequences of 0's and 1's in ultrametric analysis (see [16]).

Definition 4.6 A class φ of subsets of $\mathbb{N}$ (the set of positive integers) is said to be "ultrametric (or non-archimedean) full" if

(i) $\bigcup\limits_{S \in \varphi} S = \mathbb{N}$ (covering);

(ii) If $S \subset T$ where $T \in \varphi$, then $S \in \varphi$ (hereditary); and

(iii) If $\{t_k\}$ is a sequence in K such that $\sup\limits_{k \in S} |t_k| < \infty$ for every $S \in \varphi$, then $\sup\limits_{k \geq 1} |t_k| < \infty$.

Example 4.1 $\varphi = 2^{\mathbb{N}}$ is an example of an ultrametric full class.

Theorem 4.5 *Let φ be a class of subsets of $\mathbb{N}$ satisfying (i) and (ii) of Definition 4.6. Then φ is ultrametric full if and only if for any infinite matrix (a_{nk}) for which*

$$\sup_{n \geq 1} \left(\sup_{k \in S} |a_{nk}| \right) < \infty \text{ for every } S \in \varphi, \text{ then } \sup_{n,k \geq 1} |a_{nk}| < \infty.$$

Proof Necessity. Let φ be ultrametric full. Suppose for some infinite matrix (a_{nk}),

$\sup_{n \geq 1} \left(\sup_{k \in S} |a_{nk}| \right) < \infty$ for every $S \in \varphi$ but $\sup_{n,k \geq 1} |a_{nk}| = \infty$. We can now choose strictly increasing sequences $\{n(j)\}, \{k(j)\}$ of positive integers such that

$$M(j) = \sup_{k(j-1) < i \leq k(j)} |a_{n(j),i}| > \frac{1}{\rho^{2j}},$$

where, since K is non-trivially valued, $\pi \in K$ is such that $0 < \rho = |\pi| < 1$. Let $\mathbb{N}(j) = \{i / k(j-1) < i \leq k(j)\}$, $j = 1, 2, \ldots, k(0) = 1$. Now, define

$$b_i = a_{n(j),i} \pi^j, \quad i \in \mathbb{N}(j), \, j = 1, 2, \ldots.$$

Now,

$$\sup_{i \in \mathbb{N}(j)} |b_i| = \sup_{i \in \mathbb{N}(j)} |a_{n(j),i}| \rho^j$$
$$= \rho^j M(j)$$
$$> \rho^j \frac{1}{\rho^{2j}}$$
$$= \frac{1}{\rho^j},$$

so that

$$\sup_{i \geq 1} |b_i| = \infty,$$

since $\frac{1}{\rho^j} \to \infty$, $j \to \infty$, $\frac{1}{\rho} > 1$. Since φ is ultrametric full, there exists $S \in \varphi$ with $\sup_{i \in S} |b_i| = \infty$. Consequently, we have

$$\sup_{i \in S \cap \mathbb{N}(j)} |b_i| > 1 \text{ for infinitely many } j\text{'s},$$

for, otherwise, $\sup_{i \in S \cap \mathbb{N}(j)} |b_i| \leq 1, \, j = 1, 2, \ldots$ and so $\sup_{i \geq 1} |b_i| \leq 1$, a contradiction. Hence, for these infinitely many j's,

$$\sup_{i \in S} |a_{n(j),i}| \geq \sup_{i \in S \cap \mathbb{N}(j)} |a_{n(j),i}|$$

$$= \sup_{i \in S \cap \mathbb{N}(j)} \frac{|b_i|}{\rho^j}$$

$$> \frac{1}{\rho^j} \to \infty, j \to \infty, \text{ since} \frac{1}{\rho} > 1,$$

contrading the fact that

$$\sup_{n \geq 1} \left(\sup_{k \in S} |a_{nk}| \right) < \infty \text{ for every } S \in \varphi.$$

Sufficiency. Let $\{t_k\}$ be any sequence in K such that $\sup_{k \in S} |t_k| < \infty$ for every $S \in \varphi$. Define the matrix (a_{nk}), where $a_{nk} = t_k$, $k = 1, 2, \ldots$; $n = 1, 2, \ldots$. Then $\sup_{n \geq 1} \left(\sup_{k \in S} |a_{nk}| \right) < \infty$ for every $S \in \varphi$. By hypothesis, $\sup_{n,k \geq 1} |a_{nk}| < \infty$. It now follows that $\sup_{k \geq 1} |t_k| < \infty$ and so φ is ultrametric full, completing the proof of the theorem. $\square$

Corollary 4.1 φ *is a class of subsets of* $\mathbb{N}$ *satisfying* (i) *and* (ii) *of Definition 4.6. Then* φ *is ultrametric full if and only if for any infinite matrix* (a_{nk}) *for which*

$$\sup_{n \geq 1} \left| \sum_{k \in S} a_{nk} \right| < \infty \text{ for every } S \in \varphi, \text{ then } \sup_{n,k \geq 1} |a_{nk}| < \infty.$$

Proof Necessity. Let φ be ultrametric full. Let (a_{nk}) be an infinite matrix for which $\sup_{n \geq 1} \left| \sum_{k \in S} a_{nk} \right| < \infty$ for every $S \in \varphi$. Let $S \in \varphi$ and $k_0 \in S$. Since φ is hereditary, $S' = S \setminus \{k_0\} \in \varphi$. So

$$\sup_{n \geq 1} \left| \sum_{k \in S} a_{nk} - \sum_{k \in S'} a_{nk} \right| < \infty,$$

$$\text{i.e. } \sup_{n \geq 1} |a_{nk_0}| < \infty,$$

for every $k_0 \in S$ and so $\sup_{n \geq 1} \left(\sup_{k \in S} |a_{nk}| \right) < \infty$ for every $S \in \varphi$. Since φ is ultrametric full, it follows from Theorem 4.5 that $\sup_{n,k \geq 1} |a_{nk}| < \infty$.

Sufficiency. Let (a_{nk}) be an infinite matrix such that $\sup_{n \geq 1} \left(\sum_{k \in S} |a_{nk}| \right) < \infty$ for every $S \in \varphi$. Then

$$\sup_{n \geq 1} \left| \sum_{k \in S} a_{nk} \right| \leq \sup_{n \geq 1} \left(\sup_{k \in S} |a_{nk}| \right)$$

$$< \infty$$

for every $S \in \varphi$. By hypothesis, $\sup_{n,k \geq 1} |a_{nk}| < \infty$ and so φ is ultrametric full, using Theorem 4.5, completing the proof. $\square$

The following result is worthwhile to record.

Theorem 4.6 *There is no minimal ultrametric full class.*

Proof Let S_0 be any infinite subset of an ultrametric full class φ and

$$\Delta = \{S \in \varphi / S_0 \not\subseteq S\}.$$

Then $\Delta \subsetneq \varphi$ and Δ satisfies (i), (ii) of Definition 4.6. Let $\{t_k\}$ be a sequence in K such that $\sup_{k \geq 1} |t_k| = \infty$. Since φ is ultrametric full, there exists $W \in \varphi$ such that $\sup_{k \in W} |t_k| = \infty$. So $\sup_{k \in W \backslash S_0} |t_k| = \infty$ or $\sup_{k \in W \cap S_0} |t_k| = \infty$. In the first case, if $T = W \backslash S_0$, then $T \in \Delta$ and $\sup_{k \in T} |t_k| = \infty$. In the second case, let $T = S_0 \backslash \{s\}$, where $s \in S_0$. Then $T \in \Delta$ and $\sup_{k \in T} |t_k| \geq \sup_{k \in W \cap S_0} |t_k| = \infty$. In view of Definition 4.6, Δ is ultrametric full, where $\Delta \subsetneq \varphi$. Thus there is no minimal ultrametric full class. $\square$

We define $\chi_\varphi = \{\chi_S / S \in \varphi\}$, where χ_S denotes the characteristic function of the subset S of $\mathbb{N}$. C_A denotes the set of all sequences $x = \{x_k\}$ which are A-summable.

As an application to matrix summability, we have the following result.

Theorem 4.7 *Let φ be an ultrametric full class and $A = (a_{nk})$ be any infinite matrix. Then $C_A \supseteq \chi_\varphi$ if and only if*

(i) $\lim_{k \to \infty} a_{nk} = 0$, $n = 1, 2, \ldots$;

and

(ii) $\lim_{n \to \infty} \sup_{k \in S} |a_{n+1,k} - a_{nk}| = 0$ *for every $S \in \varphi$.*

Proof Necessity. Let $C_A \supseteq \chi_\varphi$. It is clear that (i) holds. So

$$\lim_{k \to \infty} (a_{n+1,k} - a_{nk}) = 0.$$

Suppose (ii) does not hold. We use the "sliding hump method" to arrive at a contradiction. We can now choose $\epsilon > 0$, $S \in \varphi$ and two strictly increasing sequences $\{n(i)\}$ and $\{k(i)\}$ of positive integers such that

$$\sup_{k \in S} |a_{n(i)+1,k} - a_{n(i),k}| > \epsilon;$$

$$\sup_{1 \le k \le k(i-1)} |a_{n(i)+1,k} - a_{n(i),k}| < \frac{\epsilon}{8};$$

and

$$\sup_{k > k(i)} |a_{n(i)+1,k} - a_{n(i),k}| < \frac{\epsilon}{8}.$$

In view of the above inequalities, there exists $k(n(i)) \in S, k(i-1) < k(n(i)) \le k(i)$ such that

$$|a_{n(i)+1,k(n(i))} - a_{n(i),k(n(i))}| > \epsilon.$$

Define $x = \{x_k\}$, where

$$x_k = \begin{cases} 1, & \text{if } k = k(n(i)); \\ 0, & \text{otherwise.} \end{cases}$$

Now,

$$(Ax)_{n(i)+1} - (Ax)_{n(i)} = \sum_{k=1}^{\infty} \{a_{n(i)+1,k} - a_{n(i),k}\} x_k$$

$$= \sum_{k=1}^{k(i-1)} \{a_{n(i)+1,k} - a_{n(i),k}\} x_k$$

$$+ \sum_{k=k(i-1)+1}^{k(i)} \{a_{n(i)+1,k} - a_{n(i),k}\} x_k$$

$$+ \sum_{k=k(i)+1}^{\infty} \{a_{n(i)+1,k} - a_{n(i),k}\} x_k$$

$$= \sum_{k=1}^{k(i-1)} \{a_{n(i)+1,k} - a_{n(i),k}\} x_k$$

$$+ \{a_{n(i)+1,k(n(i))} - a_{n(i),k(n(i))}\}$$

$$+ \sum_{k=k(i)+1}^{\infty} \{a_{n(i)+1,k} - a_{n(i),k}\} x_k$$

so that

$$\epsilon < |a_{n(i)+1,k(n(i))} - a_{n(i),k(n(i))}|$$
$$\le \max \left[|(Ax)_{n(i)+1} - (Ax)_{n(i)}|, \frac{\epsilon}{8}, \frac{\epsilon}{8} \right]$$

which implies that

$$|(Ax)_{n(i)+1} - (Ax)_{n(i)}| > \epsilon, \quad i = 1, 2, \ldots.$$

Thus $x \notin C_A$. Note, however, that $x \in \chi_\varphi$. Consequently, $\chi_\varphi \nsubseteq C_A$, a contradiction. So (ii) holds.

Sufficiency. Let (i), (ii) hold. In view of (i), $\sum_{k \in S} a_{nk}$ converges for every $S \in \varphi$.

Now,

$$\left| \sum_{k \in S} a_{n+1,k} - \sum_{k \in S} a_{nk} \right| = \left| \sum_{k \in S} (a_{n+1,k} - a_{nk}) \right|$$

$$\leq \sup_{k \in S} |a_{n+1,k} - a_{nk}|$$

$$\to 0, n \to \infty, \text{ using } (ii),$$

which implies that $\lim_{n \to \infty} \sum_{k \in S} a_{nk}$ exists for every $S \in \varphi$. Thus $C_A \supseteq \chi_\varphi$, completing the proof of the theorem. $\quad\square$

Corollary 4.2 (Hahn's theorem for the ultrametric case) *An infinite matrix $A = (a_{nk})$ sums all bounded sequences if and only if it sums all sequences of 0's and 1's.*

Proof Leaving out the trivial part of the result, suppose A sums all sequences of 0's and 1's, i.e. $C_A \supseteq \chi_\varphi$, where $\phi = 2^{\mathbb{N}}$. Since $\mathbb{N} \in \varphi$,

$$\lim_{k \to \infty} a_{nk} = 0, \quad n = 1, 2, \ldots.$$

Also,

$$\lim_{n \to \infty} \sup_{k \in \mathbb{N}} |a_{n+1,k} - a_{nk}| = 0,$$

i.e. $\lim_{n \to \infty} \sup_{k \geq 1} |a_{n+1,k} - a_{nk}| = 0.$

In view of Theorem 2 of [9], it follows that A sums all bounded sequences, completing the proof. $\quad\square$

In the context of $\overline{\left\{ \bigcup_{r=1}^{\infty} \Lambda_r \right\}}$ (i.e. closure of $\bigcup_{r=1}^{\infty} \Lambda_r$ in ℓ_∞), we introduce the notion of "generalized semiperiodic sequences".

Definition 4.7 $x = \{x_k\}$ is called a "generalized semiperiodic sequence", if for any $\epsilon > 0$, there exist $n, k_0 \in \mathbb{N}$ such that

$$|x_k - x_{k+sn}| < \epsilon, \quad k \geq k_0, s = 0, 1, 2, \ldots.$$

Let Q denote the set of all generalized semiperiodic sequences. One can prove that Q is a closed linear subspace of ℓ_∞. Further, whatever be K,

$$Q \subseteq \overline{\left\{ \bigcup_{r=1}^\infty \Lambda_r \right\}}.$$

When K is a complete, non-trivially valued, ultrametric field,

$$Q = \overline{\left\{ \bigcup_{r=1}^\infty \Lambda_r \right\}}.$$

Whatever be K, we shall define for $\alpha > 0$,

$$\ell_\alpha = \left\{ x = \{x_k\},\, x_k \in K,\, k = 0, 1, 2, \ldots,\, \sum_{k=0}^\infty |x_k|^\alpha < \infty \right\}.$$

One of the interesting results of ultrametric analysis is the characterization of infinite matrices in $(\ell_\alpha, \ell_\alpha)$, $\alpha > 0$ as given by the following theorem.

Theorem 4.8 (see [17], Theorem 2.1) *If K is a complete, non-trivially valued, ultrametric field and $A = (a_{nk})$, $a_{nk} \in K$, $n, k = 0, 1, 2, \ldots$, then $A \in (\ell_\alpha, \ell_\alpha)$, $\alpha > 0$, if and only if*

$$\sup_{k \geq 0} \sum_{n=0}^\infty |a_{nk}|^\alpha < \infty. \tag{4.18}$$

Proof Since $|\cdot|$ is a non-archimedean valuation, we first note that

$$\left| |a|^\alpha - |b|^\alpha \right| \leq |a + b|^\alpha \leq |a|^\alpha + |b|^\alpha, \quad \alpha > 0. \tag{4.19}$$

Sufficiency. If $x = \{x_k\} \in \ell_\alpha$, $\sum_{k=0}^\infty a_{nk} x_k$ converges, $n = 0, 1, 2, \ldots$, since $x_k \to 0$, $k \to \infty$ and $\sup_{n,k} |a_{nk}| < \infty$ by (4.18). Also,

$$\sum_{n=0}^\infty |(Ax)_n|^\alpha \leq \sum_{n=0}^\infty \sum_{k=0}^\infty |a_{nk}|^\alpha |x_k|^\alpha$$

$$\leq \left(\sum_{k=0}^\infty |x_k|^\alpha \right) \left(\sup_{k \geq 0} \sum_{n=0}^\infty |a_{nk}|^\alpha \right)$$

$$< \infty,$$

so that $\{(Ax)_n\} \in \ell_\alpha$.

Necessity. Suppose $A \in (\ell_\alpha, \ell_\alpha)$. We first note that $\sup_{k \geq 0} |a_{nk}|^\alpha = B_n < \infty$, $n = 0, 1, 2, \ldots$. For, if for some m, $\sup_{k \geq 0} |a_{mk}|^\alpha = B_m = \infty$, then, we can choose a strictly increasing sequence $\{k(i)\}$ of positive integers such that $|a_{m,k(i)}|^\alpha > i^2$, $i = 1, 2, \ldots$. Define the sequence $\{x_k\}$, where

$$x_k = \begin{cases} \frac{1}{a_{m,k}}, & k = k(i) \\ 0, & k \neq k(i) \end{cases}, i = 1, 2, \ldots.$$

Then

$$\{x_k\} \in \ell_\alpha, \text{ for, } \sum_{k=0}^\infty |x_k|^\alpha = \sum_{i=1}^\infty |x_{k(i)}|^\alpha < \sum_{i=1}^\infty \frac{1}{i^2} < \infty,$$

while $a_{m,k(i)} x_{k(i)} = 1 \nrightarrow 0, i \to \infty$, which is a contradiction. Since $(Ax)_n = a_{nk}$ for the sequence $x = \{x_n\}$, $x_n = 0, n \neq k, x_k = 1$ and $\{(Ax)_n\} \in \ell_\alpha$,

$$\mu_k = \sum_{n=0}^\infty |a_{nk}|^\alpha < \infty, \quad k = 0, 1, 2, \ldots.$$

Suppose $\{\mu_k\}$ is unbounded. Choose a positive integer $k(1)$ such that

$$\mu_{k(1)} > 3.$$

Then choose a positive integer $n(1)$ such that

$$\sum_{n=n(1)+1}^\infty |a_{n,k(1)}|^\alpha < 1,$$

so that

$$\sum_{n=0}^{n(1)} |a_{n,k(1)}|^\alpha > 2.$$

More generally, given the positive integers $k(j), n(j), j \leq m - 1$, choose positive integers $k(m)$ and $n(m)$ such that $k(m) > k(m-1), n(m) > n(m-1)$,

$$\sum_{n=n(m-2)+1}^{n(m-1)} \sum_{k=k(m)}^\infty B_n k^{-2} < 1,$$

$$\mu_{k(m)} > 2 \sum_{n=0}^{n(m-1)} B_n + \rho^{-\alpha} m^2 \left\{ 2 + \sum_{i=1}^{m-1} i^{-2} \mu_{k(i)} \right\}$$

and

$$\sum_{n=n(m)+1}^{\infty} |a_{n,k(m)}|^{\alpha} < \sum_{n=0}^{n(m-1)} B_n,$$

where, since K is non-trivially valued, there exists $\pi \in K$ such that $0 < \rho = |\pi| < 1$. Now,

$$\sum_{n=n(m-1)+1}^{n(m)} |a_{n,k(m)}|^{\alpha} = \mu_{k(m)} - \sum_{n=0}^{n(m-1)} |a_{n,k(m)}|^{\alpha} - \sum_{n=n(m)+1}^{\infty} |a_{n,k(m)}|^{\alpha}$$

$$> 2 \sum_{n=0}^{n(m-1)} B_n + \rho^{-\alpha} m^2 \left\{ 2 + \sum_{i=1}^{m-1} i^{-2} \mu_{k(i)} \right\}$$

$$- \sum_{n=0}^{n(m-1)} B_n - \sum_{n=0}^{n(m-1)} B_n$$

$$= \rho^{-\alpha} m^2 \left\{ 2 + \sum_{i=1}^{m-1} i^{-2} \mu_{k(i)} \right\}.$$

For every $i = 1, 2, \ldots$, there exists a non-negative integer $\lambda(i)$ such that

$$\rho^{\lambda(i)+1} \leq i^{-\frac{2}{\alpha}} < \rho^{\lambda(i)}.$$

Define the sequence $x = \{x_k\}$ as follows:

$$x_k = \begin{cases} \pi^{\lambda(i)+1}, & k = k(i) \\ 0, & k \neq k(i) \end{cases}, i = 1, 2, \ldots.$$

$\{x_k\} \in \ell_\alpha$, for, $\displaystyle\sum_{k=0}^{\infty} |x_k|^{\alpha} = \sum_{i=1}^{\infty} |x_{k(i)}|^{\alpha} \leq \sum_{i=1}^{\infty} \frac{1}{i^2} < \infty.$

However, using (4.19),

$$\sum_{n=n(m-1)+1}^{n(m)} |(Ax)_n|^{\alpha} \geq \Sigma_1 - \Sigma_2 - \Sigma_3,$$

where

$$\Sigma_1 = \sum_{n=n(m-1)+1}^{n(m)} |a_{n,k(m)}|^\alpha |x_{k(m)}|^\alpha,$$

$$\Sigma_2 = \sum_{n=n(m-1)+1}^{n(m)} \sum_{i=1}^{m-1} |a_{n,k(i)}|^\alpha |x_{k(i)}|^\alpha,$$

$$\Sigma_3 = \sum_{n=n(m-1)+1}^{n(m)} \sum_{i=m+1}^{\infty} |a_{n,k(i)}|^\alpha |x_{k(i)}|^\alpha.$$

Now,

$$\Sigma_1 = \sum_{n=n(m-1)+1}^{n(m)} |a_{n,k(m)}|^\alpha \rho^{(\lambda(m)+1)\alpha}$$

$$\geq \rho^\alpha \sum_{n=n(m-1)+1}^{n(m)} |a_{n,k(m)}|^\alpha m^{-2}$$

$$> 2 + \sum_{i=1}^{m-1} i^{-2} \mu_{k(i)};$$

$$\Sigma_2 = \sum_{n=n(m-1)+1}^{n(m)} \sum_{i=1}^{m-1} |a_{n,k(i)}|^\alpha \rho^{(\lambda(i)+1)\alpha}$$

$$\leq \sum_{i=1}^{m-1} i^{-2} \sum_{n=n(m-1)+1}^{n(m)} |a_{n,k(i)}|^\alpha$$

$$\leq \sum_{i=1}^{m-1} i^{-2} \mu_{k(i)};$$

$$\Sigma_3 = \sum_{n=n(m-1)+1}^{n(m)} \sum_{i=m+1}^{\infty} |a_{n,k(i)}|^\alpha \rho^{(\lambda(i)+1)\alpha}$$

$$\leq \sum_{n=n(m-1)+1}^{n(m)} \sum_{i=k(m+1)}^{\infty} B_n i^{-2}$$

$$< 1.$$

Using the above, we have

$$\sum_{n=n(m-1)+1}^{n(m)} |(Ax)_n|^\alpha > 1, \quad m = 2, 3, \ldots.$$

This shows that $\{(Ax)_n\} \notin \ell_\alpha$, while $\{x_k\} \in \ell_\alpha$, a contradiction. Thus condition (4.18) is also necessary, completing the proof of the theorem.   □

Because of the fact that there is, as such, no classical analogue for the above result, Theorem 4.8 is interesting. When $K = \mathbb{R}$ or $\mathbb{C}$, a complete characterization of the class $(\ell_\alpha, \ell_\beta)$ of infinite matrices, $\alpha, \beta \geq 2$, does not seem to be available in the literature.

Necessary and sufficient conditions for $A \in (\ell_1, \ell_1)$ when $K = \mathbb{R}$ or $\mathbb{C}$ are due to Mears [18] (for alternate proofs, see, for instance, Fridy [19]). From the characterization mentioned in Theorem 4.8, it is deduced that the Cauchy product of two sequences in ℓ_α, $\alpha > 0$, is again in ℓ_α, a result which fails to hold for $\alpha > 1$, when $K = \mathbb{R}$ or $\mathbb{C}$.

We write $A = (a_{nk}) \in (\ell_\alpha, \ell_\alpha; P)$ if $A \in (\ell_\alpha, \ell_\alpha)$ and $\sum_{n=0}^{\infty}(Ax)_n = \sum_{k=0}^{\infty} x_k$, $x = \{x_k\} \in \ell_\alpha$; $A \in (\ell_\alpha, \ell_\alpha; P)'$ if $A \in (\ell_\alpha, \ell_\alpha; P)$ and (4.16) holds. When $K = \mathbb{R}$ or $\mathbb{C}$, Fridy [20] proved a Steinhaus-type result in the form

$$(\ell_1, \ell_1; P) \cap (\ell_\alpha, \ell_1) = \phi, \quad \alpha > 1.$$

This result, as such, fails to hold in the ultrametric set up (see [17], Remark 4.1). However, in the ultrametric set up, Natarajan [17] proved that

$$(\ell_\alpha, \ell_\alpha; P)' \cap (\ell_\beta, \ell_\alpha) = \phi, \quad \beta > \alpha.$$

In the above context, it is worth noting that following the proof of Theorem 4.1 of [17], we can show that given any matrix $A \in (\ell_\alpha, \ell_\alpha; P)$, there exists a sequence of 0's and 1's whose A-transform is not in ℓ_α. This is analogous to Schur's version of the well-known Steinhaus theorem for regular matrices (see [17]).

4.3 Core of a Sequence and Knopp's Core Theorem

The core of a complex sequence is defined as follows:

Definition 4.8 If $x = \{x_k\}$ is a complex sequence, we denote by $K_n(x)$, $n = 0, 1, 2, \ldots$, the smallest closed convex set containing $x_n, x_{n+1}, \ldots$.

$$\mathscr{K}(x) = \bigcap_{n=0}^{\infty} K_n(x)$$

is defined as the core of x.

It is known [21] that if $x = \{x_k\}$ is bounded,

$$\mathscr{K}(x) = \bigcap_{z \in \mathbb{C}} C_{\overline{\lim_{n \to \infty}} |z - x_n|}(z),$$

where $C_r(z)$ is the closed ball centred at z and radius r. Sherbakoff [21] generalized the notion of the core of a bounded complex sequence by introducing the idea of the generalized α-core $\mathscr{K}^{(\alpha)}(x)$ of a bounded complex sequence as

$$\mathscr{K}^{(\alpha)}(x) = \bigcap_{z \in \mathbb{C}} C_{\alpha \overline{\lim_{n \to \infty}} |z - x_n|}(z), \quad \alpha \geq 1. \tag{4.20}$$

When $\alpha = 1$, $\mathscr{K}^{(\alpha)}(x)$ reduces to the usual core $\mathscr{K}(x)$. When $K = \mathbb{C}$, Sherbakoff [21] showed that under the condition

$$\overline{\lim_{n \to \infty}} \left(\sum_{k=0}^{\infty} |a_{nk}| \right) = \alpha, \quad \alpha \geq 1, \tag{4.21}$$

$$\mathscr{K}(A(x)) \subseteq \mathscr{K}^{(\alpha)}(x).$$

Natarajan [22] improved the result of Sherbakoff by showing that his result works with the less stringent precise condition

$$\overline{\lim_{n \to \infty}} \left(\sum_{k=0}^{\infty} |a_{nk}| \right) \leq \alpha, \quad \alpha \geq 1, \tag{4.22}$$

(4.22) being also necessary besides the regularity of A for

$$\mathscr{K}(A(x)) \subseteq \mathscr{K}^{(\alpha)}(x)$$

for any bounded complex sequence x. This result for the case $\alpha = 1$ yields a simple proof of Knopp's core theorem (see, for instance, [23]). Natarajan's theorem is as follows.

Theorem 4.9 ([22], Theorem 2.1) *When $K = \mathbb{R}$ or $\mathbb{C}$, $A = (a_{nk})$ is such that*

$$\mathscr{K}(A(x)) \subseteq \mathscr{K}^{(\alpha)}(x), \quad \alpha \geq 1,$$

for any bounded sequence x if and only if A is regular and satisfies (4.22),

$$\text{i.e.} \ \overline{\lim_{n \to \infty}} \left(\sum_{k=0}^{\infty} |a_{nk}| \right) \leq \alpha.$$

Proof Let $x = \{x_n\}$ be a bounded sequence. If $y \in \mathcal{H}(A(x))$, for any $z \in K$,

$$|y - z| \leq \overline{\lim_{n \to \infty}} |z - (Ax)_n|.$$

If A is a regular matrix satisfying (4.22), then

$$|y - z| \leq \overline{\lim_{n \to \infty}} |z - (Ax)_n|$$

$$= \overline{\lim_{n \to \infty}} \left| \sum_{k=0}^{\infty} a_{nk}(z - x_k) \right|$$

$$\leq \alpha \overline{\lim_{k \to \infty}} |z - x_k|,$$

$$\text{i.e. } y \in C_{\alpha \overline{\lim_{k \to \infty}} |z - x_k|}(z) \ \text{ for any } z \in K,$$

which implies that $\mathcal{H}(A(x)) \subseteq \mathcal{H}^{(\alpha)}(x)$.

Conversely, let $\mathcal{H}(A(x)) \subseteq \mathcal{H}^{(\alpha)}(x)$. Then it is clear that A is regular by considering convergent sequences for which

$$\mathcal{H}^{(\alpha)}(x) = \left\{ \lim_{n \to \infty} x_n \right\}.$$

It remains to prove (4.22). Let, if possible,

$$\overline{\lim_{n \to \infty}} \left(\sum_{k=0}^{\infty} |a_{nk}| \right) > \alpha.$$

Then

$$\overline{\lim_{n \to \infty}} \left(\sum_{k=0}^{\infty} |a_{nk}| \right) = \alpha + h, \ \text{ for some } h > 0.$$

Using the hypothesis and the fact that A is regular, we can now choose two strictly increasing sequences $\{n(i)\}$ and $\{k(n(i))\}$ of positive integers such that

$$\sum_{k=0}^{k(n(i-1))} |a_{n(i),k}| < \frac{h}{8},$$

$$\sum_{k=k(n(i-1))+1}^{k(n(i))} |a_{n(i),k}| > \alpha + \frac{h}{4}$$

and

$$\sum_{k=k(n(i))+1}^{\infty} |a_{n(i),k}| < \frac{h}{8}.$$

Define the sequence $x = \{x_k\}$ by

$$x_k = sgn(a_{n(i),k}), \quad k(n(i-1)) \leq k < k(n(i)), \quad i = 1, 2, \ldots.$$

Now,

$$|(Ax)_{n(i)}| \geq \sum_{k=k(n(i-1))+1}^{k(n(i))} |a_{n(i),k}| - \sum_{k=0}^{k(n(i-1))} |a_{n(i),k}| - \sum_{k=k(n(i))+1}^{\infty} |a_{n(i),k}|$$

$$> \alpha + \frac{h}{4} - \frac{h}{8} - \frac{h}{8}$$

$$= \alpha, \quad i = 1, 2, \ldots.$$

By the regularity of A, $\{(Ax)_{n(i)}\}_{i=1}^{\infty}$ is a bounded sequence. It has a convergent subsequence whose limit cannot be in $C_\alpha(0)$ in view of the above inequality, i.e. $|(Ax)_{n(i)}| > \alpha, i = 1, 2, \ldots.$ Using (4.20), we have $\mathscr{K}^{(\alpha)}(x) \subseteq C_\alpha(0)$ for the sequence x chosen above. This leads to a contradiction of the fact that $\mathscr{K}(A(x)) \subseteq \mathscr{K}^{(\alpha)}(x)$, completing the proof.   □

In [22], Natarajan also proved the analogue of Theorem 4.9 when K is a complete, locally compact, non-trivially valued, ultrametric field.

Definition 4.9 If $x = \{x_n\}$, $x \in K$, $n = 0, 1, 2, \ldots$, we denote by $K_n(x)$, $n = 0, 1, 2, \ldots$ the smallest closed K-convex set containing $x_n, x_{n+1}, \ldots$ and call

$$\mathscr{K}(x) = \bigcap_{n=0}^{\infty} K_n(x)$$

the core of x. $\mathscr{K}^{(\alpha)}(x)$ is defined as in (4.20) in this case too.

Theorem 4.10 ([22], Theorem 3.1) *K is a complete, locally compact, non-trivially valued, ultrametric field. $A = (a_{nk})$, $a_{nk} \in K$, $n, k = 0, 1, 2, \ldots$ is such that $\mathscr{K}(A(x)) \subseteq \mathscr{K}^{(\alpha)}(x)$ for any sequence (bounded or unbounded) if and only if A is regular and satisfies*

$$\overline{\lim_{n \to \infty}} \left(\sup_{k \geq 0} |a_{nk}| \right) \leq \alpha, \quad \alpha > 0. \tag{4.23}$$

Note that the above theorem, for $\alpha = 1$, yields a simple proof of the analogue of Knopp's core theorem in the ultrametric set up.

4.4 A Characterization of Regular and Schur Matrices

When $K = \mathbb{R}$ or $\mathbb{C}$, Maddox [24] obtained a characterization of Schur matrices in terms of the existence of a bounded divergent sequence all of whose subsequences are summable by the matrix. This characterization included the earlier one of Buck's [25], viz., a sequence $\{x_k\}$, summable by a regular matrix A, is convergent if and only if A sums each one of its subsequences. Fridy [26] showed that we can replace "subsequence" in Buck's result by "rearrangement" to obtain that a sequence $\{x_k\}$, summable by a regular matrix A, is convergent if and only if A sums each one of its rearrangements.

In this connection, it is to be reiterated that it is not as if every result in ultrametric analysis has a proof analogous to its classical counterpart or even a simpler proof. The absence of analogues for the signum function, upper limit and lower limit of real number sequences, etc. forces on us to search for alternate devices. These devices provide an entirely different proof of even an exact analogue of a classical theorem. For instance, the proof of the following theorem, amply, illustrates our claim.

Theorem 4.11 *Let K be a complete, non-trivially valued, ultrametric field. Then $A = (a_{nk})$, $a_{nk} \in K$, $n, k = 0, 1, 2, \ldots$ is a Schur matrix if and only if there exists a bounded divergent sequence $x = \{x_k\}$, $x_k \in K$, $k = 0, 1, 2, \ldots$, each one of whose subsequences is summable A.*

Proof If A is a Schur matrix, the assertion of the theorem is a consequence of the definition of such a matrix. Conversely, let $x = \{x_k\}$ be a divergent sequence, each one of whose subsequences is summable A. For each $p = 0, 1, 2, \ldots$, we can choose two subsequences of x (say) $\{x_k^{(1)}\}$, $\{x_k^{(2)}\}$ such that if $y_k = x_k^{(1)} - x_k^{(2)}$, $y_k = 0$, $k \neq p$ while $y_p \neq 0$. Such choice is possible since x diverges and therefore has two unequal entries after any stage for k. The sequence $\{y_k\}$ is summable A. So it follows that $\lim_{n \to \infty} a_{np}$ exists, $p = 0, 1, 2, \ldots$. Next, we show that $\lim_{p \to \infty} a_{np} = 0$, $n = 0, 1, 2, \ldots$. Otherwise, there exists $\epsilon' > 0$ and a non-negative integer m such that

$$|a_{m,p(i)}| > \epsilon', \ i = 1, 2, \ldots,$$

where $\{p(i)\}$ is an increasing sequence of positive integers. Since x diverges, it is not a null sequence and so there exists $\epsilon'' > 0$ and an increasing sequence $\{\ell(j)\}$ of positive integers such that

$$|x_{\ell(j)}| > \epsilon'', \ j = 1, 2, \ldots.$$

Now,

$$|a_{m,p(i)} x_{\ell(p(i))}| > \epsilon^2, \ i = 1, 2, \ldots,$$

where $\epsilon = \min(\epsilon', \epsilon'')$. This means that the A-transform of the subsequence $\{x_{\ell(j)}\}$ does not exist, a contradiction. Thus $\lim_{p \to \infty} a_{np} = 0, n = 0, 1, 2, \ldots$. Since x diverges,

$|x_{k+1} - x_k| \nrightarrow 0, k \to \infty$ so that there exists $\epsilon''' > 0$ and an increasing sequence $\{k(j)\}$ of positive integers such that

$$|x_{k(j)+1} - x_{k(j)}| > \epsilon''', \quad j = 1, 2, \ldots. \tag{4.24}$$

We may assume that $k(j+1) - k(j) > 1, j = 1, 2, \ldots$. We claim that if A is not a Schur matrix, then we should have an $\epsilon > 0$ and two increasing sequences $\{n(i)\}$ and $\{p(n(i))\}$ of positive integers with

$$\begin{cases} \text{(i)} \quad \sup_{0 \leq p \leq p(n(i-1))} |a_{n(i)+1,p} - a_{n(i),p}| < \dfrac{\epsilon^2}{4M}, \\[2mm] \text{(ii)} \ |a_{n(i)+1,p(n(i))} - a_{n(i),p(n(i))}| > \epsilon, \\[2mm] \text{(iii)} \quad \sup_{p \geq p(n(i+1))} |a_{n(i)+1,p} - a_{n(i),p}| < \dfrac{\epsilon^2}{4M}, \end{cases} \tag{4.25}$$

where $M = \sup\limits_{k \geq 0} |x_k|$. Before proving the claim, we show that if A is not a Schur matrix, then x is necessarily bounded under the hypothesis of the theorem. Suppose x is unbounded. We consider two cases:

Case (i) If A is such that $a_{nk} \neq 0$ for some n and $k = k(i), i = 1, 2, \ldots$, choose a subsequence $\{x_{\alpha(k)}\}$ of x, which is unbounded, such that

$$|a_{nk} x_{\alpha(k)}| > 1, \quad k = k(i), \ i = 1, 2, \ldots.$$

Hence, $\{x_{\alpha(k)}\}$ is not summable A, a contradiction.

Case (ii) If now, $a_{nk} = 0, k > k(n), n = 0, 1, 2, \ldots$, A not being a Schur matrix, we can find two strictly increasing sequences of positive integers $\{n(j)\}$ and $\{k(j)\}$ such that $a_{n(j),k(j)}$ is the last non-zero term in the $n(j)$th row. x, being unbounded, we can choose a subsequence $\{x_{\alpha(k)}\}$ of x such that

$$|A_{n(j)}(\{x_{\alpha(k)}\})| > j, \quad j = 1, 2, \ldots.$$

To do this, choose $x_{\alpha(i)}, i = 1, 2, \ldots, k(1)$ such that

$$|x_{\alpha(1)}| > \frac{1}{|a_{n(1),1}|}, \quad \text{if } a_{n(1),1} \neq 0,$$

while $x_{\alpha(1)}$ is chosen as a suitable x_k, otherwise. Having chosen $x_{\alpha(1)}$, choose $x_{\alpha(2)}$, $\alpha(2) > \alpha(1)$ such that

$$|x_{\alpha(2)}| > \frac{1}{|a_{n(1),2}|} \{1 + |a_{n(1),1} x_{\alpha(1)}|\}, \quad \text{if } a_{n(1),2} \neq 0.$$

Otherwise, choose $x_{\alpha(2)}$ such that $\alpha(2) > \alpha(1)$. Now,

$$
\begin{aligned}
&|a_{n(1),1}x_{\alpha(1)} + a_{n(1),2}x_{\alpha(2)}| \\
&\geq |a_{n(1),2}x_{\alpha(2)}| - |a_{n(1),1}x_{\alpha(1)}|, \quad \text{if } a_{n(1),2} \neq 0; \\
&= |a_{n(1),1}x_{\alpha(1)}|, \quad \text{if } a_{n(1),2} \neq 0.
\end{aligned}
$$

Thus

$$
\begin{aligned}
&|a_{n(1),1}x_{\alpha(1)} + a_{n(1),2}x_{\alpha(2)}| \\
&= 0, \quad \text{if } a_{n(1),1} = 0 = a_{n(1),2}; \\
&> 1, \quad \text{if one of } a_{n(1),1}, a_{n(1),2} \neq 0.
\end{aligned}
$$

We choose $x_{\alpha(i)}, i = 1, 2, \ldots, k(1)$ as above. Then

$$
\left| \sum_{k=1}^{k(1)} a_{n(1),k}x_{\alpha(k)} \right| > 1.
$$

If now,

$$
\sum_{k=1}^{k(1)} a_{n(2),k}x_{\alpha(k)} = \alpha,
$$

choose similarly $x_{\alpha(k)}$, $k(1) < k \leq k(2)$ with $\alpha(k(1)) < \alpha(k(1)+1) < \cdots < \alpha(k(2))$ and

$$
\left| \sum_{k=k(1)+1}^{k(2)} a_{n(2),k}x_{\alpha(k)} \right| > 2 + |\alpha|.
$$

Then,

$$
\begin{aligned}
\left| \sum_{k=1}^{k(2)} a_{n(2),k}x_{\alpha(k)} \right| &\geq \left| \sum_{k=k(1)+1}^{k(2)} a_{n(2),k}x_{\alpha(k)} \right| - \left| \sum_{k=1}^{k(1)} a_{n(2),k}x_{\alpha(k)} \right| \\
&> 2 + |\alpha| - |\alpha| = 2.
\end{aligned}
$$

Inductively, we can therefore choose $x_{\alpha(k)}$, $k = 1, 2, \ldots$ with

$$
|A_{n(j)}(\{x_{\alpha(k)}\})| = \left| \sum_{k=1}^{k(j)} a_{n(j),k}x_{\alpha(k)} \right|
$$

$$
> j, \quad j = 1, 2, \ldots.
$$

It now follows that $\{x_{\alpha(k)}\}$ is not summable A, a contradiction. Thus in both cases it turns out that x is bounded, if A were not to be a Schur matrix.

Next, we observe that since A is not a Schur matrix (see [9]), there exist $\epsilon > 0$ and an increasing sequence $\{n(i)\}$ of positive integers such that

$$\sup_{p \geq 0} |a_{n(i)+1,p} - a_{n(i),p}| > \epsilon, \quad i = 1, 2, \ldots.$$

Hence, there exists $p(n(i))$ such that

$$|a_{n(i)+1,p(n(i))} - a_{n(i),p(n(i))}| > \epsilon, \quad i = 1, 2, \ldots. \tag{4.26}$$

Suppose $\{p(n(i))\}$ is bounded, then there are only a finite number of distinct entries in that sequence. Consequently, there exists $p = p(n(m))$ which occurs in the sequence $\{p(n(i))\}$ infinite number of times. For this p, (4.26) will then contradict the existence of $\lim_{n \to \infty} a_{np}$, $p = 0, 1, 2, \ldots$ established earlier. Having chosen $\{n(i)\}$ and $\{p(n(i))\}$ to satisfy (4.26), it is clear that by choosing a subsequence of $\{n(i)\}$, if necessary, we can suppose that (4.25) holds. Consider now the sequence $\{y_p\}$ defined as follows:

$$y_p = \begin{cases} x_{k(p)+1}, & p = p(n(i)), \\ x_{k(p)}, & p \neq p(n(i)), i = 1, 2, \ldots, \end{cases}$$

where the sequence $\{k(j)\}$ is already chosen as in (4.24). Thus

$$\left| \sum_{p=0}^{\infty} \{a_{n(i)+1,p} - a_{n(i),p}\}(y_p - x_{k(p)}) \right|$$

$$> \epsilon^2 - \frac{\epsilon^2}{4M}M - \frac{\epsilon^2}{4M}M, \text{ using (4.25)}$$

$$= \frac{\epsilon^2}{2}, \quad i = 1, 2, \ldots,$$

where we can suppose that $\epsilon \geq \epsilon'''$. This contradicts the fact that

$$\left\{ \sum_{p=0}^{\infty} (a_{n+1,p} - a_{n,p})(y_p - x_{k(p)}) \right\}_{n=0}^{\infty}$$

converges. This proves that A is a Schur matrix. The proof of the theorem is now complete.   □

The analogue of Buck's result in the ultrametric case follows as a corollary of Theorem 4.11, viz., a sequence $x = \{x_k\}$, $x_k \in K$, $k = 0, 1, 2, \ldots$, summable by a regular matrix A, is convergent if and only if every one of its subsequences is summable A.

The following analogue in the ultrametric case can also be established by means of a "sliding hump method" as described by Fridy [26].

Theorem 4.12 ([28], Theorem 3) *A sequence* $x = \{x_k\}$, $x_k \in K$, $k = 0, 1, 2, \ldots$, *summable by a regular matrix A, is convergent if and only if every one of its rearrangements is summable A.*

When $K = \mathbb{R}$ or $\mathbb{C}$, in the context of rearrangements of a bounded sequence, Fridy [26] proved the following result.

Theorem 4.13 *A null sequence* $x = \{x_k\}$ *is in* ℓ_1 *if and only if there exists a matrix* $A = (a_{nk}) \in (\ell_1, \ell_1; P)$ *which transforms all rearrangements of* $\{x_k\}$ *into sequences in* ℓ_1.

We can combine Theorem 4.13 and a result of Keagy [27] to state

Theorem 4.14 *When* $K = \mathbb{R}$ *or* $\mathbb{C}$, *a null sequence* $x = \{x_k\}$ *is in* ℓ_1 *if and only if there exists a matrix* $A = (a_{nk}) \in (\ell_1, \ell_1; P)$ *which transforms every subsequence or rearrangement of x into a sequence in* ℓ_1.

In the ultrametric setting, Theorem 4.14 fails (see [28]). However, when K is a complete, non-trivially valued, ultrametric field, the following theorem, due to Natarajan ([28], Theorem 6), is an attempt to salvage Fridy's result in a general form.

Theorem 4.15 *A null sequence* $x = \{x_k\}$ *is in* ℓ_α *if and only if there exists a matrix* $A = (a_{nk}) \in (\ell_\alpha, \ell_\alpha; P)'$ *such that A transforms every rearrangement of x into a sequence in* ℓ_α, *where we recall that* $A \in (\ell_\alpha, \ell_\alpha; P)'$ *if* $A \in (\ell_\alpha, \ell_\alpha; P)$ *and* (4.16) *holds.*

Proof We recall (see [17]) that $A \in (\ell_\alpha, \ell_\alpha; P)$ if and only if (4.18) holds and

$$\sum_{n=0}^{\infty} a_{nk} = 1, \quad k = 0, 1, 2, \ldots.$$

Leaving out the trivial part of the theorem, suppose $x \in c_0 - \ell_\alpha$ and $A \in (\ell_\alpha, \ell_\alpha; P)'$ transforms every rearrangement of x into a sequence in ℓ_α. Choose $k(1) = 1$ and a positive integer $n(1)$ such that

$$\sum_{n=n(1)+1}^{\infty} |a_{n,1}|^\alpha < 2^{-1},$$

so that

$$\sum_{n=0}^{n(1)} |a_{n,1}|^\alpha = \sum_{n=0}^{\infty} |a_{n,1}|^\alpha - \sum_{n=n(1)+1}^{\infty} |a_{n,1}|^\alpha$$

$$\geq 1 - \frac{1}{2} = \frac{1}{2}.$$

Having defined $k(j), n(j), j \leq m-1$, choose a positive integer $k(m) > k(m-1)+1$ such that

$$\sum_{n=0}^{n(m-1)} |a_{n,k(m)}|^\alpha < 2^{-m}, \quad |x_{k(m)}| < \frac{1}{m^{\frac{2}{\alpha}}}$$

and then choose a positive integer $n(m) > n(m-1)$ such that

$$\sum_{n=n(m-1)+1}^{n(m)} |a_{n,k(m)}|^\alpha \geq \frac{1}{2},$$

$$\sum_{n=n(m)+1}^{\infty} |a_{n,k(m)}|^\alpha < 2^{-m}.$$

Let U consist of all $k(m)$, $m = 1, 2, \ldots$. Let $u_m = x_{k(m)}$ and V be the set of all non-negative integers which are not in U. Let $v = \{x_k\}_{k \in V}$. Let y be a rearrangement of x, where

$$y_k = v_m, \quad k = k(m);$$
$$= u_k, \quad \text{otherwise.}$$

Defining $n(0) = 0$, we have

$$
\begin{aligned}
\sum_{n=0}^{n(M)} |(Ay)_n|^\alpha &\geq \sum_{m=1}^{M} \sum_{n=n(m-1)+1}^{n(m)} \left\{ \left| \sum_{k \in U} a_{nk} y_k \right|^\alpha \right. \\
&\quad \left. - \left| \sum_{k \in V} a_{nk} y_k \right|^\alpha \right\}, \quad \text{using (4.19)} \\
&\geq \sum_{m=1}^{M} \sum_{n=n(m-1)+1}^{n(m)} \left\{ \left| \sum_{k \in U} a_{nk} y_k \right|^\alpha - \sum_{k \in V} |a_{nk} y_k|^\alpha \right\} \\
&\geq \sum_{m=1}^{M} \sum_{n=n(m-1)+1}^{n(m)} \left\{ |a_{n,k(m)} v_m|^\alpha - \sum_{\substack{i=1 \\ i \neq m}} |a_{n,k(i)} v_i|^\alpha \right. \\
&\quad \left. - \sum_{k \in V} |a_{nk} y_k|^\alpha \right\} \\
&\geq \frac{1}{2} \sum_{m=1}^{M} |v_m|^\alpha - \sum_{m=1}^{M} \sum_{n=n(m-1)+1}^{n(m)} \sum_{i=1}^{\infty} |a_{n,k(i)} v_i|^\alpha \\
&\quad - \sum_{n=0}^{n(M)} \sum_{k \in V} |a_{nk} y_k|^\alpha.
\end{aligned}
\tag{4.27}
$$

However,

$$\sum_{m=1}^{M} \sum_{n=n(m-1)+1}^{n(m)} \sum_{\substack{i=1 \\ i \neq m}}^{\infty} |a_{n,k(i)} v_i|^{\alpha} < ||x||^{\alpha} \sum_{m=1}^{\infty} 2^{-m+1}, \tag{4.28}$$

where $||x|| = \sup_{k \geq 0} |x_k|$, for,

$$\sum_{m=1}^{\infty} \sum_{n=n(m-1)+1}^{n(m)} \sum_{i<m} |a_{n,k(i)}|^{\alpha}$$

$$= \sum_{m=1}^{\infty} \sum_{n=n(m)+1}^{\infty} |a_{n,k(m)}|^{\alpha}$$

$$< \sum_{m=1}^{\infty} 2^{-m},$$

and similarly

$$\sum_{m=1}^{\infty} \sum_{n=n(m-1)+1}^{n(m)} \sum_{i>m} |a_{n,k(i)}|^{\alpha} < \sum_{m=1}^{\infty} 2^{-(m+1)}.$$

Also,

$$\left[\begin{aligned} \sum_{n=0}^{n(M)} \sum_{k \in V} |a_{nk} y_k|^{\alpha} &= \sum_{k \in V} \sum_{n=0}^{n(M)} |a_{nk} u_k|^{\alpha} \\ &\leq \left(\sup_{k \geq 0} \sum_{n=0}^{\infty} |a_{nk}|^{\alpha} \right) \sum_{k \in V} |u_k|^{\alpha} \\ &< \infty. \end{aligned} \right. \tag{4.29}$$

In view of (4.27)–(4.29),

$$\sum_{n=0}^{n(M)} |(Ay)_n|^{\alpha} \to \infty, \quad M \to \infty,$$

since $u = \{u_k\} \in \ell_{\alpha}$ and so $v \notin \ell_{\alpha}$, i.e. $Ay \notin \ell_{\alpha}$, a contradiction, completing the proof of the theorem.   □

4.5 Cauchy Multiplication of Series

Let c_0 denote the set of all null sequences with entries in K. With respect to the norm in ℓ_∞, c_0 is a closed subspace of the ultrametric space ℓ_∞. The effective analogue in the ultrametric set up of the classical space ℓ_1 seems to be c_0. Equivalently, absolute convergence in classical analysis is effectively replaced by ordinary convergence in ultrametric analysis. For instance, the following theorem is a violent departure from classical theory and it is very useful in the sequel.

Theorem 4.16 ([29], Theorem 1) *If* $\displaystyle\sum_{k=0}^{\infty} a_k$ *and* $\displaystyle\sum_{k=0}^{\infty} b_k$ *are two infinite series, then*

$\displaystyle\sum_{k=0}^{\infty} c_k$ *converges for every convergent series* $\displaystyle\sum_{k=0}^{\infty} a_k$ *if and only if* $\displaystyle\sum_{k=0}^{\infty} b_k$ *converges, where*

$$c_k = a_0 b_k + a_1 b_{k-1} + \cdots + a_k b_0, \quad k = 0, 1, 2, \ldots .$$

In other words, for every sequence $\{a_n\}$ *with* $\lim_{n\to\infty} a_n = 0$, $\lim_{n\to\infty} c_n = 0$ *if and only if* $\lim_{n\to\infty} b_n = 0$.

If $\displaystyle\sum_{k=0}^{\infty} a_k$ *and* $\displaystyle\sum_{k=0}^{\infty} b_k$ *both converge, then* $\displaystyle\sum_{k=0}^{\infty} c_k$ *converges and*

$$\sum_{k=0}^{\infty} c_k = \left(\sum_{k=0}^{\infty} a_k \right) \left(\sum_{k=0}^{\infty} b_k \right).$$

Proof Let $\displaystyle\sum_{k=0}^{\infty} b_k$ be given such that $\displaystyle\sum_{k=0}^{\infty} c_k$ converges for every convergent series $\displaystyle\sum_{k=0}^{\infty} a_k$. For the series $\displaystyle\sum_{k=0}^{\infty} a_k$, $a_0 = 1$, $a_k = 0$, $k = 1, 2, \ldots$, $c_k = b_k$ so that $\displaystyle\sum_{k=0}^{\infty} b_k$ converges. Conversely, let $\displaystyle\sum_{k=0}^{\infty} a_k$ and $\displaystyle\sum_{k=0}^{\infty} b_k$ converge. There exists $M > 0$ such that $|a_k| < M$, $|b_k| < M$, $k = 0, 1, 2, \ldots$. Given $\epsilon > 0$, choose a positive integer N_1 such that $|a_k| < \frac{\epsilon}{M}$, for all $k > N_1$. Since $\lim_{k\to\infty} b_{k-r} = 0$, $r = 0, 1, 2, \ldots$, we can choose a positive integer $N_2 > N_1$ such that for $k > N_2$, $\sup_{0 \le r \le N_1} |b_{k-r}| < \frac{\epsilon}{M}$. Thus for $k > N_2$,

$$|c_k| = \left| \sum_{r=0}^{k} b_{k-r} a_r \right|$$

$$= \left| \sum_{r=0}^{N_1} b_{k-r} a_r + \sum_{r=N+1}^{k} b_{k-r} a_r \right|$$

$$\leq \max \left\{ \max_{0 \leq r \leq N_1} |b_{k-r}||a_r|, \ \max_{N_1+1 \leq r \leq k} |b_{k-r}||a_r| \right\}$$

$$< \max \left\{ \left(\frac{\epsilon}{M}\right) M, M \left(\frac{\epsilon}{M}\right) \right\}$$

$$= \epsilon.$$

Consequently, $\lim\limits_{k \to \infty} c_k = 0$ and so $\sum\limits_{k=0}^{\infty} c_k$ converges. It is easy to check that in this case

$$\sum_{k=0}^{\infty} c_k = \left(\sum_{k=0}^{\infty} a_k \right) \left(\sum_{k=0}^{\infty} b_k \right),$$

completing the proof.   □

References

1. Andree, R.V., Petersen, G.M.: Matrix methods of summation, regular for p-adic valuations. Proc. Amer. Math. Soc. **7**, 250–253 (1956)
2. Roberts, J.B.: Matrix summability in F-fields. Proc. Amer. Math. Soc. **8**, 541–543 (1957)
3. Monna, A.F.: Sur le théorème de Banach-Steinhaus. Indag. Math. **25**, 121–131 (1963)
4. Rangachari, M.S., Srinivasan, V.K.: Matrix transformations in non-archimedean fields. Indag. Math. **26**, 422–429 (1964)
5. Srinivasan, V.K.: On certain summation processes in the p-adic field. Indag. Math. **27**, 319–325 (1965)
6. Somasundaram, D.: Some properties of T-matrices over non-archimedean fields. Publ. Math. Debrecen **21**, 171–177 (1974)
7. Somasundaram, D.: On a theorem of Brudno over non-archimedean fields. Bull. Austral. Math. Soc. **23**, 191–194 (1981)
8. Natarajan, P.N.: Criterion for regular matrices in non-archimedean fields. J. Ramanujan Math. Soc. **6**, 185–195 (1991)
9. Natarajan, P.N.: The Steinhaus theorem for Toeplitz matrices in non-archimedean fields. Comment. Math. Prace Mat. **20**, 417–422 (1978)
10. Natarajan, P.N.: A Steinhaus type theorem. Proc. Amer. Math. Soc. **99**, 559–562 (1987)
11. Natarajan, P.N.: Some Steinhaus type theorems over valued fields. Ann. Math. Blaise Pascal **3**, 183–188 (1996)
12. Natarajan, P.N.: Some more Steinhaus type theorems over valued fields. Ann. Math. Blaise Pascal **6**, 47–54 (1999)
13. Natarajan, P.N.: Some more Steinhaus type theorems over valued fields II. Comm. Math. Analysis **5**, 1–7 (2008)
14. Natarajan, P.N.: On certain spaces containing the space of Cauchy sequences. J. Orissa Math. Soc. **9**, 1–9 (1990)
15. Sember, J.J., Freedman, A.R.: On summing sequences of 0's and 1's. Rocky Mountain J. Math. **11**, 419–425 (1981)

16. Natarajan, P.N.: On sequences of zeros and ones in non-archimedean analysis—A further study. Afr. Diaspora J. Math. **10**, 49–54 (2010)
17. Natarajan, P.N.: Characterization of a class of infinite matrices with applications. Bull. Austral. Math. Soc. **34**, 161–175 (1986)
18. Mears, F.M.: Absolute regularity and the Nörlund mean. Ann. Math. **38**, 594–601 (1937)
19. Fridy, J.A.: A note on absolute summability. Proc. Amer. Math. Soc. **20**, 285–286 (1969)
20. Fridy, J.A.: Properties of absolute summability matrices. Proc. Amer. Math. Soc. **24**, 583–585 (1970)
21. Sherbakoff, A.A.: On cores of complex sequences and their regular transform (Russian). Mat. Zametki **22**, 815–828 (1977)
22. Natarajan, P.N.: On the core of a sequence over valued fields. J. Indian. Math. Soc. **55**, 189–198 (1990)
23. Cooke, R.G.: Infinite Matrices and Sequence Spaces. Macmillan, London (1950)
24. Maddox, I.J.: A Tauberian theorem for subsequences. Bull. London Math. Soc. **2**, 63–65 (1970)
25. Buck, R.C.: A note on subsequences. Bull. Amer. Math. Soc. **49**, 898–899 (1943)
26. Fridy, J.A.: Summability of rearrangements of sequences. Math. Z. **143**, 187–192 (1975)
27. Keagy, T.A.: Matrix transformations and absolute summability. Pacific J. Math. **63**, 411–415 (1976)
28. Natarajan, P.N.: Characterization of regular and Schur matrices over non-archimedean fields. Proc. Kon. Ned. Akad. Wetens Ser. A **90**, 423–430 (1987)
29. Natarajan, P.N.: Multiplication of series with terms in a non-archimedean field. Simon Stevin **52**, 157–160 (1978)

Chapter 5
The Nörlund and The Weighted Mean Methods

Abstract In this chapter, we introduce the Nörlund and the Weighted Mean methods in the ultrametric set-up and their properties are elaborately discussed. We also show that the Mazur–Orlicz theorem and Brudno's theorem fail to hold in the ultrametric case.

Keywords The Nörlund method · The weighted mean method · Mazur–Orlicz theorem · Brudno's theorem · Translative (matrix)

5.1 The Nörlund Method

In an attempt to introduce special summability methods in ultrametric analysis, Srinivasan [1] defined the Nörlund method of summability, i.e. the (N, p_n) method in K as follows: The (N, p_n) method is defined by the infinite matrix (a_{nk}) where

$$a_{nk} = \frac{p_{n-k}}{P_n}, \quad k \leq n;$$
$$= 0, \quad k > n,$$

where $|p_0| > |p_j|, j = 1, 2, \ldots,$ and $P_n = \sum_{k=0}^{n} p_k, n = 0, 1, 2, \ldots$. It is to be noted that $p_0 \neq 0$. Srinivasan noted that if

$$\lim_{n \to \infty} p_n = 0, \tag{5.1}$$

then the (N, p_n) method is regular. An example of a regular (N, p_n) method, in the p-adic field $\mathbb{Q}_p$, for a prime p, is given by $p_n = p^n, n = 0, 1, 2, \ldots$. We now observe that (5.1) is also necessary for the (N, p_n) method to be regular. Thus we have

© Springer India 2015
P.N. Natarajan, *An Introduction to Ultrametric Summability Theory*,
Forum for Interdisciplinary Mathematics 2, DOI 10.1007/978-81-322-2559-1_5

Theorem 5.1 ([2], Theorem 1) *The method* (N, p_n) *is regular if and only if* (5.1) *holds.*

Proof If (5.1) holds, the method is regular (see [1], p. 323). Conversely, if the method is regular, then $\lim_{n \to \infty} a_{n0} = 0$. But $a_{n0} = \frac{p_n}{P_n}$ so that $|a_{n0}| = \frac{|p_n|}{|p_0|}$ (since $|P_n| = |p_0|$ using Theorem 1.1), which implies that (5.1) holds.   □

We now prove a limitation theorem for the (N, p_n) method.

Theorem 5.2 ([2], Lemma 3) *If* $\{s_n\}$ *is* (N, p_n) *summable, then* $\{s_n\}$ *is bounded.*

Proof Let $\{t_n\}$ be the (N, p_n) transform of $\{s_k\}$, i.e.

$$t_n = \frac{p_0 s_n + p_1 s_{n-1} + \cdots + p_n s_0}{P_n}, \quad n = 0, 1, 2, \ldots .$$

Since $\{t_n\}$ converges, let $M = \sup_{n \geq 0} |t_n|$. Now,

$$t_0 = \frac{p_0 s_0}{P_0} = \frac{p_0 s_0}{p_0} = s_0,$$

and so $|s_0| \leq M$. Let $|s_k| \leq M$, $k = 0, 1, \ldots, (n-1)$. Since

$$s_n = \frac{P_n t_n - p_1 s_{n-1} - \cdots - p_n s_0}{p_0},$$

$$|s_n| \leq \frac{|p_0| M}{|p_0|} = M,$$

completing the proof.   □

Definition 5.1 We write $(N, p_n) \subseteq (N, q_n)$ if whenever $\{s_n\}$ is summable (N, p_n) to s then it is also summable (N, q_n) to s. We say that (N, p_n) and (N, q_n) are equivalent if $(N, p_n) \subseteq (N, q_n)$ and vice versa.

The following two results are essentially different from their counterparts in the classical case (see [3], p. 67, Theorem 2.1). It is worth noting in this context that these results are also instances in which absolute convergence in classical analysis is replaced by ordinary convergence in ultrametric analysis.

Theorem 5.3 ([2], Theorem 3) *Let* (N, p_n), (N, q_n) *be regular methods. Then* $(N, p_n) \subseteq (N, q_n)$ *if and only if* $\lim_{n \to \infty} k_n = 0$, *where* $\frac{q(x)}{p(x)} = k(x) = \sum_{n=0}^{\infty} k_n x^n$,

$p(x) = \sum_{n=0}^{\infty} p_n x^n$, $q(x) = \sum_{n=0}^{\infty} q_n x^n$, *and* $\{k_n\}$ *is defined by* $k_n p_0 + k_{n-1} p_1 + \cdots + k_0 p_n = q_n$, $n = 0, 1, 2, \ldots$ *by recursion.*

Proof Let $(N, p_n) \subseteq (N, q_n)$. Let $\{t_n\}$, $\{\tau_n\}$ be, respectively, the (N, p_n), (N, q_n) transforms of the sequence $\{s_n\}$. If $|x| < 1$,

$$\sum_{n=0}^{\infty} Q_n \tau_n x^n = \sum_{n=0}^{\infty} Q_n \left(\sum_{k=0}^{n} \frac{q_{n-k}}{Q_n} s_k \right) x^n$$

$$= \sum_{n=0}^{\infty} (q_0 s_n + q_1 s_{n-1} + \cdots + q_n s_0) x^n$$

$$= q(x) s(x).$$

Similarly,

$$\sum_{n=0}^{\infty} P_n t_n x^n = p(x) s(x); \quad \text{if } |x| < 1.$$

Now,

$$k(x) p(x) = q(x),$$
$$k(x) p(x) s(x) = q(x) s(x),$$
$$\text{i.e. } k(x) \left(\sum_{n=0}^{\infty} P_n t_n x^n \right) = \sum_{n=0}^{\infty} Q_n \tau_n x^n.$$

Thus

$$Q_n \tau_n = P_n t_n k_0 + P_{n-1} t_{n-1} k_1 + \cdots + P_0 t_0 k_n,$$

$$\text{i.e. } \tau_n = \sum_{j=0}^{\infty} a_{nj} t_j,$$

where

$$a_{nj} = \begin{cases} \frac{k_{n-j} P_j}{Q_n}, & j \le n; \\ 0, & j > n. \end{cases}$$

By hypothesis, (a_{nj}) is a regular matrix and so $\lim_{n \to \infty} a_{n0} = 0$. Thus

$$0 = \lim_{n \to \infty} |a_{n0}| = \lim_{n \to \infty} \frac{|k_n||p_0|}{|q_0|},$$

which, in turn, implies that $k_n \to 0$, $n \to \infty$.
Conversely, let $k_n \to 0$, $n \to \infty$. For $j = 0, 1, 2, \ldots$,

$$\lim_{n \to \infty} |a_{nj}| = \lim_{n \to \infty} \frac{|k_{n-j}||P_j|}{|Q_n|}$$

$$= \lim_{n \to \infty} \frac{|k_{n-j}||p_0|}{|q_0|}$$

$$= 0.$$

$|a_{nj}| = \frac{|k_{n-j}||p_0|}{|q_0|} \leq M', \; M' = \frac{L|p_0|}{|q_0|}, \; L = \sup_{n \geq 0} |k_n|, \; n, j = 0, 1, 2, \ldots$ so that $\sup_{n,j} |a_{nj}| < \infty$. Finally,

$$\sum_{j=0}^{\infty} a_{nj} = \sum_{j=0}^{n} a_{nj} = \frac{k_0 P_n + k_1 P_{n-1} + \cdots + k_n P_0}{Q_n}$$

$$= \frac{Q_n}{Q_n} = 1, \quad n = 0, 1, 2, \ldots,$$

so that $\lim_{n \to \infty} \sum_{j=0}^{\infty} a_{nj} = 1$. Thus (a_{nj}) is regular and consequently $(N, p_n) \subseteq (N, q_n)$, completing the proof. $\square$

Theorem 5.4 ([2], Theorem 4) *The regular methods (N, p_n), (N, q_n) are equivalent if and only if $\lim_{n \to \infty} k_n = 0 = \lim_{n \to \infty} h_n$, where $\{k_n\}$ is defined as in Theorem 5.3 and $\{h_n\}$ is defined by* $\dfrac{p(x)}{q(x)} = h(x) = \displaystyle\sum_{n=0}^{\infty} h_n x^n.$

The following results too are different from their classical counterparts just like Theorems 5.3 and 5.4.

Theorem 5.5 ([4], Theorem 1) *If $a_k = o(1)$, $k \to \infty$, i.e. $\lim_{k \to \infty} a_k = 0$, $\{b_k\}$ is summable by a regular (N, p_n) method, then $\{c_k\}$ is (N, p_n) summable, where*

$$c_k = \sum_{i=0}^{k} a_i b_{k-i}, \; k = 0, 1, 2, \ldots.$$

Theorem 5.6 ([4], Theorem 2) *If $\displaystyle\sum_{k=0}^{\infty} a_k$ converges to A, $\displaystyle\sum_{k=0}^{\infty} b_k$ is summable to B by a regular (N, p_n) method, then $\displaystyle\sum_{k=0}^{\infty} c_k$ is (N, p_n) summable to AB, where c_k is defined as in Theorem 5.5.*

In the classical case, the following result, due to Mears, is known ([5], Theorem 1).

Theorem 5.7 *If the real series $\displaystyle\sum_{k=0}^{\infty} a_k$, $\displaystyle\sum_{k=0}^{\infty} b_k$ are such that the former is summable by regular (N, p_n) method to A, the latter is summable by a regular (N, q_n) method* *to B, one of the summabilities being absolute, i.e.* $\sum\limits_{n=0}^{\infty} |t_{n+1} - t_n| < \infty$, *where* $\{t_n\}$

is the (N, p_n) *transform of* $\{A_k\}$ *or the* (N, q_n) *transform of* $\{B_k\}$, $A_k = \sum\limits_{j=0}^{k} a_j$,

$B_k = \sum\limits_{j=0}^{k} b_j$, $k = 0, 1, 2, \ldots$, *then* $\sum\limits_{k=0}^{\infty} c_k$ *is summable by the* (N, r_n) *method to* AB,

where $c_k = \sum\limits_{j=0}^{k} a_j b_{k-j}$, $r_k = \sum\limits_{j=0}^{k} p_j q_{k-j}$, $k = 0, 1, 2, \ldots$.

In the ultrametric case, we have the following analogue of Theorem 5.7, which indicates significant departure in the sense that there is no need to bring in absolute summability (see [5], Theorem 2).

Theorem 5.8 *If* $\sum\limits_{k=0}^{\infty} a_k$ *is summable by a regular* (N, p_n) *method to* A, $\sum\limits_{k=0}^{\infty} b_k$ *is*

summable by a regular (N, q_n) *method to* B, *then* $\sum\limits_{k=0}^{\infty} c_k$ *is summable by the regular*

(N, r_n) *method to* AB, *where* $c_k = \sum\limits_{j=0}^{k} a_j b_{k-j}$, $r_k = \sum\limits_{j=0}^{k} p_j q_{k-j}$, $k = 0, 1, 2, \ldots$.

In the context of Theorem 5.8, we note that the hypothesis that "$(N, p_n), (N, q_n)$ are regular for $\sum\limits_{k=0}^{\infty} c_k$ to be summable (N, r_n) to AB" cannot be dropped (see [6], pp. 50–52).

However, we have the following result in which $(N, p_n), (N, q_n)$ need not be regular.

Theorem 5.9 ([6], Theorem 1) *Let* $\sum\limits_{k=0}^{\infty} a_k$ *be* (N, p_n) *summable to 0 and* $\sum\limits_{k=0}^{\infty} b_k$ *be*

(N, q_n) *summable to 0. Then* $\sum\limits_{k=0}^{\infty} c_k$ *is* (N, r_n) *summable to 0.*

There is a natural isomorphism φ from c_0 onto the ultrametric algebra A of power series converging in the closed unit disc $B(0, 1) = \{x \in K / |x| \leq 1\}$, defined by

$$\varphi(\{a_n\}) = \sum\limits_{n=0}^{\infty} a_n x^n$$

(For instance, see [7]). Both c_0 and A are ultrametric Banach spaces with respect to the sup norm, i.e. $\|\{a_n\}\| = \sup_{n \geq 0} |a_n|$. If $f = \varphi(\{a_n\})$, then $f(1) = \sum_{n=0}^{\infty} a_n$. Given sequences $\{a_n\}, \{b_n\} \in c_0$, we know that their Cauchy product $\{c_n\} \in c_0$, by Theorem 4.16. Putting $f = \varphi(\{a_n\})$, $g = \varphi(\{b_n\})$, $h = \varphi(\{c_n\})$, we have $h(x) = f(x)g(x)$. Let $M = (m_{nk}) \in (c_0, c_0)$, i.e. $\sup_{n,k} |m_{nk}| < \infty$ and $\lim_{n \to \infty} m_{nk} = 0$, $k = 0, 1, 2, \ldots$. Let $\{a_n\} \in c_0$ and $\{\alpha_n\} = M(\{a_n\})$, i.e. $\{\alpha_n\}$ is the M-transform of $\{a_n\}$. We then know that $\{\alpha_n\} \in c_0$. Thus M defines a linear operator $\tilde{M}$ of A, defined by

$$\tilde{M}\left(\sum_{n=0}^{\infty} a_n x^n \right) = \sum_{n=0}^{\infty} \alpha_n x^n.$$

This applies to regular Nörlund matrices in particular.

In view of the above remarks, Theorem 5.9 brings out the following connection between regular Nörlund matrices and analytic functions in the closed unit disc.

Theorem 5.10 (see [6], Theorem 2) *Let M, N be regular Nörlund matrices and $f, g \in A$. If $\tilde{M}f(1) = \tilde{N}g(1) = 0$, then we have $\tilde{M} \circ \tilde{N}(fg)(1) = 0$.*

Definition 5.2 The convergence field (or summability field) c_A of any infinite matrix $A = (a_{nk})$ is defined as

$$c_A = \{x = \{x_k\} : Ax = \{(Ax)_n\} \in c\}.$$

In [8], we prove two interesting results regarding the convergence field of a regular Nörlund method.

Theorem 5.11 *Given a regular Nörlund method (N, p_n), the convergence field of (N, p_n) is c if and only if*

$$p(z) = \sum_{n=0}^{\infty} p_n z^n \neq 0, \quad on \ |z| \leq 1.$$

Theorem 5.12 *There is no $p = \{p_n\}$ such that the convergence field of (N, p_n) is ℓ_∞.*

The following result is well known in classical summability theory (see [9], p. 231).

Theorem 5.13 (Mazur–Orlicz) *If a conservative matrix sums a bounded divergent sequence, it sums an unbounded one.*

The above Theorem 5.13 fails to hold in ultrametric analysis in view of Theorem 5.2.

In [10], Natarajan and Sakthivel proved some more results regarding the convergence fields of regular Nörlund means in the ultrametric case. For instance, the following theorems were proved.

Theorem 5.14 ([10], Lemma 3.5) *Consider a function* $p(x) = \sum_{n=0}^{\infty} p_n x^n$ *regular for* $|x| < 1$ *and a function* $r(x) = \sum_{n=0}^{\infty} r_n x^n$ *regular for* $|x| \leq 1$. *Define* $q(x)$ *by* $q(x) = r(x)p(x)$. *If* (N, p) *is regular, then* (N, q) *is regular and*

$$0((N, p)) = 0((N, q)),$$

where (N, p) *denotes the Nörlund method* (N, p_n), (N, q) *denotes the Nörlund method* (N, q_n) *and* $0((N, p))$ *denotes the set of all sequences which are* (N, p) *summable to 0 with similar meaning for* $0((N, q))$.

The following result brings out an additive relation between convergence fields of Nörlund means.

Theorem 5.15 ([10], Theorem 4.3) *If* $r(x) \neq 0$ *for* $|x| = 1$ *and* $p(x) \neq 0$ *for* $|x| < 1$, *then* $\{s_n\} \in 0((N, q))$ *if and only if*

$$s_n = u_n + v_n,$$

where $\{u_n\} \in 0((N, r))$ *and* $\{v_n\} \in 0((N, p))$.

Definition 5.3 A conservative matrix $A = (a_{nk})$ for which $\sum_{k=0}^{\infty} \left(\lim_{n \to \infty} a_{nk} \right) = \lim_{n \to \infty} \left(\sum_{k=0}^{\infty} a_{nk} \right)$ is said to be "conull".

Wilansky ([11] p. 34) proved the following theorem in the classical case.

Theorem 5.16 *Every conservative, non-regular real Nörlund matrix is conull.*

However, in the ultrametric case, we have the following result, which is a violent departure from the classical case.

Theorem 5.17 ([12], Theorem 2) *Any conservative, non-regular Nörlund matrix is never conull.*

Corollary 5.1 *Since any Schur matrix is conull (see [13], p. 160), it follows that any conservative, non-regular Nörlund matrix is never a Schur matrix.*

5.2 Some More Properties of the Nörlund Method

We now record some interesting properties of regular Nörlund method (see [14]).
The following result is easily proved.

Theorem 5.18 $A = (a_{nk}) \in (c_0, c_0)$ *if and only if (4.1) and (4.2) hold with* $\delta_k = 0$, $k = 0, 1, 2, \ldots$.

Consequently, we have

Theorem 5.19 *The Nörlund method* $(N, p) \in (c_0, c_0)$ *if and only if* $p = \{p_n\} \in c_0$.

Theorem 5.20 ([14], Theorem 2.3) *The following statements are equivalent:*

(i) (N, p) *is regular;*
(ii) $(N, p) \in (c_0, c_0)$;
(iii) $p \in c_0$;
and
(iv) $\lim\limits_{n \to \infty} P_n = P \neq 0$.

Let N_{c_0} denote the set of all Nörlund methods that are (c_0, c_0), i.e. N_{c_0} denotes the set of all regular Nörlund methods. Let $c_0((N, p))$ denote the set of all sequences $x = \{x_k\}$ such that $(N, p)(x) \in c_0$.

Definition 5.4 Given Nörlund methods $(N, p), (N, q) \in N_{c_0}$, we say that $(N, g) = (N, p * q)$ is the symmetric product of (N, p) and (N, q) provided $(N, g) \in \mathcal{N}$, where $\mathcal{N}$ denotes the set of all Nörlund methods and $g = \{g_n\} = p * q$ is the Cauchy (symmetric) product of the sequences $p = \{p_n\}$ and $q = \{q_n\}$, i.e. $g_n = \sum\limits_{k=0}^{n} p_k q_{n-k}$, $n = 0, 1, 2, \ldots$.

Lemma 5.1 ([14], Lemma 2.1) *Let* $p = \{p_n\}, q = \{q_n\}$ *be two sequences in* K *and let* $r = p * q$. *Suppose* $(N, p), (N, r) \in \mathcal{N}$. *Then*

$$c_0((N, p)) \subseteq c_0((N, r))$$

if and only if $(N, q) \in N_{c_0}$.

Proof For any sequence $x = \{x_k\}$ in K,

$$((N, r)x)_n = \frac{1}{R_n} \sum_{k=0}^{n} q_{n-k} P_k((N, p)x)_k$$

$$= \sum_{k=0}^{n} \ell_{nk}((N, p)x)_k,$$

where

$$\ell_{nk} = \begin{cases} \frac{q_{n-k}P_k}{R_n}, & k \le n; \\ 0, & k > n. \end{cases}$$

It is easy to show that $|R_n| = |r_0|$, $n = 0, 1, 2, \ldots$. Now, if $k \le n$,

$$|\ell_{nk}| = \left| \frac{q_{n-k}P_k}{R_n} \right| \le \frac{|q_0||p_0|}{|r_0|};$$

if $k > n$, $\ell_{nk} = 0$. Thus $\sup_{n,k} |\ell_{nk}| < \infty$. In view of Theorem 5.18,

$$c_0((N, p)) \subseteq c_0((N, r))$$

if and only if

$$\lim_{n \to \infty} \ell_{nk} = 0, \quad k = 0, 1, 2, \ldots,$$

i.e. $$\lim_{n \to \infty} \left| \frac{q_{n-k}P_k}{R_n} \right| = 0, \quad k = 0, 1, 2, \ldots,$$

i.e. $$\lim_{n \to \infty} \frac{|p_0|}{|r_0|} |q_{n-k}| = 0, \quad k = 0, 1, 2, \ldots,$$

i.e. $$\lim_{n \to \infty} q_{n-k} = 0, \quad k = 0, 1, 2, \ldots,$$

i.e. $$\lim_{n \to \infty} q_n = 0.$$

In other words, $(N, q) \in N_{c_0}$, completing the proof. □

The following result is an immediate consequence of the above lemma.

Theorem 5.21 ([14], Theorem 2.7) *Suppose* $(N, p), (N, q) \in N_{c_0}$. *Then*

$$c_0((N, p)) \subseteq c_0((N, q))$$

if and only if $b = \{b_n\} \in c_0$, *which is defined by*

$$b(x) = \frac{q(x)}{p(x)} = \sum_{n=0}^{\infty} b_n x^n,$$

$$p(x) = \sum_{n=0}^{\infty} p_n x^n, \ q(x) = \sum_{n=0}^{\infty} q_n x^n.$$

Proof In the above Lemma 5.1, replace q by b so that

$$r = p * b = q.$$

The result now follows.   □

Corollary 5.2 *Let* $(N, p), (N, q) \in N_{c_0}$. *Then*

(i) $c_0((N, p)) = c_0((N, q))$ *if and only if both* $a = \{a_n\} \in c_0$ *and* $b = \{b_n\} \in c_0$, *where* $\{a_n\}$ *is defined by*

$$a(x) = \frac{p(x)}{q(x)} = \sum_{n=0}^{\infty} a_n x^n;$$

(ii) $c_0((N, p)) \subsetneqq c_0((N, q))$ *if and only if both* $a = \{a_n\} \notin c_0$ *and* $b = \{b_n\} \in c_0$.

Corollary 5.3 *Let* $(N, p) \in N_{c_0}$ *and* $h(x) = \frac{1}{p(x)}$. *Then* $c_0((N, p)) = c_0$ *if and only if* $h \in c_0$.

Proof Let I be the identity matrix so that $c_0(I) = c_0$. Then $I(x) = \sum_{n=0}^{\infty} i_n x^n, i_0 = 1$, $i_n = 0, n = 1, 2, \ldots$. So $a(x) = \frac{p(x)}{I(x)} = p(x)$ and $b(x) = \frac{I(x)}{p(x)} = \frac{1}{p(x)} = h(x)$. The conclusion now follows.   □

Corollary 5.4 *Let* $(N, p), (N, q) \in N_{c_0}$ *and* $r = p * q$. *Then*

$$c_0((N, p)) \subseteq c_0((N, r)).$$

The following result is easily established.

Theorem 5.22 ([14], Theorem 2.8) *Let* $(N, p) \in N_{c_0}$. *Let* $q = \{p'_0, p_1, p_2, \ldots\}$, *with* $|p_0| < |p'_0|$. *Then* $(N, q) \in N_{c_0}$ *and further*

$$c_0((N, p)) \cap c_0((N, q)) = c_0.$$

Natarajan proved that N_{c_0} is an ordered abelian semigroup, the order relation is the set inclusion between summability fields of type $c_0((N, p))$ and the binary operation is the Cauchy product or symmetric product of sequences. He also proved that there are infinite chains of Nörlund methods from N_{c_0} (for details, refer to [14], Sect. 3). This is similar to the Cesáro methods $(C, k), k = 0, 1, 2, \ldots$ in the classical case, which form an infinite chain.

5.3 The Weighted Mean Method

In developing summability methods in ultrametric fields, Srinivasan [1] defined the analogue of the classical weighted means $(\bar{N}, p_n)$ under the assumption that the sequence $\{p_n\}$ of weights satisfies the conditions:

$$|p_0| < |p_1| < |p_2| < \cdots < |p_n| < \cdots; \tag{5.2}$$

and

$$\lim_{n \to \infty} |p_n| = \infty. \tag{5.3}$$

However, it turned out that these weighted means were equivalent to usual convergence. Natarajan [15] remedied the situation by assuming that the sequence $\{p_n\}$ of weights satisfies the conditions:

$$p_n \neq 0, \quad n = 0, 1, 2, \ldots; \tag{5.4}$$

and

$$|p_i| \leq |P_j|, \quad i = 0, 1, 2, \ldots, j; \ j = 0, 1, 2, \ldots, \tag{5.5}$$

where $P_j = \sum_{k=0}^{j} p_k, j = 0, 1, 2, \ldots$.

Note that (5.5) is equivalent to

$$\max_{0 \leq i \leq j} |p_i| \leq |P_j|, \quad j = 0, 1, 2, \ldots.$$

Since $|\cdot|$ is an ultrametric valuation,

$$|P_j| \leq \max_{0 \leq i \leq j} |p_i|,$$

so that (5.5) is equivalent to

$$|P_j| = \max_{0 \leq i \leq j} |p_i|. \tag{5.6}$$

Definition 5.5 [15] The weighted mean method in K, denoted by $(\bar{N}, p_n)$, is defined by the infinite matrix (a_{nk}), where

$$a_{nk} = \begin{cases} \frac{p_k}{P_n}, & k \leq n; \\ 0, & k > n, \end{cases}$$

$p_n \neq 0, n = 0, 1, 2, \ldots$ and $|p_i| \leq |P_j|, i = 0, 1, 2, \ldots, j; \ j = 0, 1, 2, \ldots,$

$P_n = \sum_{k=0}^{n} p_k, n = 0, 1, 2, \ldots$.

Remark 5.1 (see [15], p. 193, Remark 1) If $\left|\frac{P_{n+1}}{P_n}\right| > 1$, $n = 0, 1, 2, \ldots$ and $\lim\limits_{n \to \infty} |P_n| = \infty$, i.e. $\{|P_n|\}$ strictly increases to infinity, then the method $(\bar{N}, p_n)$ is trivial. For, $|p_n| = |P_n - P_{n-1}| = |P_n|$, since $|P_n| > |P_{n-1}|$ and $|\cdot|$ is an ultra-metric valuation. So (5.2) and (5.3) are satisfied. Consequently, $(\bar{N}, p_n)$ is trivial, i.e. $(\bar{N}, p_n)$ is equivalent to convergence.

Remark 5.2 (see [15], p. 193, Remark 2) We note that (5.5) is equivalent to

$$|P_{n+1}| \geq |P_n|, \quad n = 0, 1, 2, \ldots. \tag{5.7}$$

Proof Let (5.5) hold. Now,

$$\begin{aligned}
|P_{n+1}| &= \max_{0 \leq i \leq n+1} |p_i| \\
&= \max \left[\max_{0 \leq i \leq n} |p_i|, |p_{n+1}| \right] \\
&= \max[|P_n|, |p_{n+1}|] \\
&\geq |P_n|, \quad n = 0, 1, 2, \ldots,
\end{aligned}$$

so that (5.7) holds. Conversely, let (5.7) hold. For a fixed integer $j \geq 0$, let $0 \leq i \leq j$. Then

$$\begin{aligned}
|p_i| &= |P_i - P_{i-1}| \\
&\leq \max[|P_i|, |P_{i-1}|] \\
&= |P_i| \\
&\leq |P_j|,
\end{aligned}$$

so that (5.5) holds.   □

It is now easy to prove the following result.

Theorem 5.23 (see [15], p. 194, Theorem 1) $(\bar{N}, p_n)$ *is regular if and only if*

$$\lim_{n \to \infty} |P_n| = \infty. \tag{5.8}$$

Example 5.1 (see [15], p. 194, Remark 4) There are non-trivial $(\bar{N}, p_n)$ methods. Let $\alpha \in K$ such that $0 < |\alpha| < 1$, this being possible since K is non-trivially valued. Let $\{p_n\} = \left\{\alpha, \frac{1}{\alpha^2}, \alpha^3, \frac{1}{\alpha^4}, \ldots\right\}$ and $\{s_n\} = \left\{\frac{1}{\alpha}, \alpha^2, \frac{1}{\alpha^3}, \alpha^4, \ldots\right\}$. It is clear that $\{s_n\}$ does not converge. However, $\{s_n\}$ is $(\bar{N}, p_n)$ summable to zero.

Theorem 5.24 (Limitation theorem) (see [15], p. 195, Theorem 2) *If* $\{s_n\}$ *is* $(\bar{N}, p_n)$ *summable to* s, *then*

$$|s_n - s| = o\left(\frac{P_n}{p_n}\right), \quad n \to \infty, \tag{5.9}$$

Proof Let $t_n = \dfrac{p_0 s_0 + p_1 s_1 + \cdots + p_n s_n}{P_n}$, $n = 0, 1, 2, \ldots$. Then, by hypothesis, $\lim\limits_{n \to \infty} t_n = s$. Now,

$$P_n t_n - P_{n-1} t_{n-1} = p_n s_n,$$

so that

$$
\begin{aligned}
\frac{p_n}{P_n}(s_n - s) &= \frac{p_n}{P_n}\left[\frac{P_n t_n - P_{n-1} t_{n-1}}{p_n} - s\right] \\
&= \frac{1}{P_n}[P_n t_n - P_{n-1} t_{n-1} - p_n s] \\
&= \frac{1}{P_n}[P_n t_n - P_{n-1} t_{n-1} - (P_n - P_{n-1})s] \\
&= \frac{1}{P_n}[P_n(t_n - s) - P_{n-1}(t_{n-1} - s)] \\
&= (t_n - s) - \frac{P_{n-1}}{P_n}(t_{n-1} - s).
\end{aligned}
$$

Consequently,

$$\left|\frac{p_n}{P_n}(s_n - s)\right| \le \max(|t_n - s|, |t_{n-1} - s|), \quad \text{since} \quad \left|\frac{P_{n-1}}{P_n}\right| \le 1$$

$$\to 0, \quad n \to \infty.$$

Thus

$$|s_n - s| = o\left(\frac{P_n}{p_n}\right), \quad n \to \infty,$$

completing the proof of the theorem. □

In [15], some interesting inclusion theorems involving weighted means were proved. We shall record them here without proofs.

Theorem 5.25 ([15], p. 195, Theorem 3) *[Comparison theorem for two weighted means] If $(\bar{N}, p_n)$, $(\bar{N}, q_n)$ are two regular weighted mean methods and if*

$$\left|\frac{P_n}{p_n}\right| \le H\left|\frac{Q_n}{q_n}\right|, \quad n = 0, 1, 2, \ldots, \tag{5.10}$$

where $H > 0$ is a constant, $P_n = \displaystyle\sum_{k=0}^{n} p_k$, $Q_n = \displaystyle\sum_{k=0}^{n} q_k$, $n = 0, 1, 2, \ldots$, then $(\bar{N}, p_n) \subseteq (\bar{N}, q_n)$.

Theorem 5.26 ([15], p. 196, Theorem 4) *[Comparison theorem for a regular* $(\bar{N}, p_n)$ *method and a regular matrix A] Let* $(\bar{N}, p_n)$ *be a regular weighted mean method and* $A = (a_{nk})$ *be a regular matrix. If*

$$\lim_{k \to \infty} \frac{a_{nk} P_k}{p_k} = 0, \quad n = 0, 1, 2, \ldots; \tag{5.11}$$

and

$$\sup_{n,k} \left| \left(\frac{a_{nk}}{p_k} - \frac{a_{n,k+1}}{p_{k+1}} \right) P_k \right| < \infty, \tag{5.12}$$

then $(\bar{N}, p_n) \subseteq A$.

Theorem 5.27 ([15], p. 197, Theorem 5) *Let* $(\bar{N}, p_n)$ *be a regular weighted mean method and* $A = (a_{nk})$ *be a regular triangular matrix. Then* $(\bar{N}, p_n) \subseteq A$ *if and only if (5.12) holds.*

In the context of Theorem 5.17, we have the following interesting result about weighted mean methods:

Theorem 5.28 ([12], p. 431, Theorem 3) *Every conservative, non-regular weighted mean matrix is a Schur matrix and hence conull.*

In view of Steinhaus theorem (Theorem 4.3), we can reformulate the above theorem as follows:

Theorem 5.29 ([12], p. 432, Theorem 4) *A conservative weighted mean matrix is non-regular if and only if it is a Schur matrix.*

Theorem 5.30 ([12], p. 432, Theorem 5) *Let* $(\bar{N}, p_n)$, $(\bar{N}, q_n)$ *be two weighted mean methods. If both of them sum the same bounded sequences and if one of them is regular, then the other is regular too.*

Definition 5.6 Infinite matrices $A = (a_{nk})$, $B = (b_{nk})$ are said to be "consistent" if $\lim_{n \to \infty} (Ax)_n = \lim_{n \to \infty} (Bx)_n$ whenever $x \in c_A \cap c_B$. B is said to be stronger than A if $c_A \subseteq c_B$.

Note that $A \subseteq B$ if and only if B is stronger than A and consistent with A.

As in the classical case ([11], p. 12), we can prove the following result in the ultrametric case too.

Theorem 5.31 ([16], Theorem 5) *Let* A, B *be triangular matrices. Then* B *is stronger than* A, *i.e.* $c_A \subseteq c_B$ *if and only if* BA^{-1} *is conservative. Also* $A \subseteq B$ *if and only* BA^{-1} *is regular.*

Brudno's result in the classical case (see [17], p. 130) is:

Theorem 5.32 *If every bounded sequence which is summable by a regular matrix A is also summable by a regular matrix B, then A and B are consistent for these sequences.*

Brudno's theorem fails to hold in the ultrametric case as the following counterexample using weighted means shows. In the p-adic field $\mathbb{Q}_p$, let

$$Y = \begin{pmatrix} 1 & 0 & 0 & 0 & 0 & \cdots \\ \frac{1}{2} & \frac{1}{2} & 0 & 0 & 0 & \cdots \\ 0 & \frac{1}{2} & \frac{1}{2} & 0 & 0 & \cdots \\ 0 & 0 & \frac{1}{2} & \frac{1}{2} & 0 & \cdots \\ & & \cdots & & & \end{pmatrix},$$

$$\bar{N} = \begin{pmatrix} 1 & 0 & 0 & 0 & 0 & \cdots \\ \frac{1}{1+p} & \frac{p}{1+p} & 0 & 0 & 0 & \cdots \\ \frac{1}{1+p+\frac{1}{p^2}} & \frac{p}{1+p+\frac{1}{p^2}} & \frac{\frac{1}{p^2}}{1+p+\frac{1}{p^2}} & 0 & 0 & \cdots \\ \cdots & \cdots & \cdots & & & \end{pmatrix}.$$

Simple computation shows that

$$Y^{-1} = \begin{pmatrix} 1 & 0 & 0 & 0 & 0 & 0 & \cdots \\ -1 & 2 & 0 & 0 & 0 & 0 & \cdots \\ 1 & -2 & 2 & 0 & 0 & 0 & \cdots \\ -1 & 2 & -2 & 2 & 0 & 0 & \cdots \\ 1 & -2 & 2 & -2 & 2 & 0 & \cdots \\ & & \cdots & & & & \end{pmatrix}$$

and

$$NY^{-1} = (a_{nk})$$

$$= \begin{pmatrix} 1 & 0 & 0 & 0 & 0 & \cdots \\ \frac{1-p}{1+p} & \frac{2p}{1+p} & 0 & 0 & 0 & \cdots \\ \frac{1-p+\frac{1}{p^2}}{1+p+\frac{1}{p^2}} & \frac{2\left(p-\frac{1}{p^2}\right)}{1+p+\frac{1}{p^2}} & \frac{2\frac{1}{p^2}}{1+p+\frac{1}{p^2}} & 0 & 0 & \cdots \\ \frac{1-p+\frac{1}{p^2}-p^3}{1+p+\frac{1}{p^2}+p^3} & \frac{2\left(p-\frac{1}{p^2}+p^3\right)}{1+p+\frac{1}{p^2}+p^3} & \frac{2\left(\frac{1}{p^2}-p^3\right)}{1+p+\frac{1}{p^2}+p^3} & \frac{2p^3}{1+p+\frac{1}{p^2}+p^3} & 0 & \cdots \\ \cdots & \cdots & \cdots & \cdots & & \end{pmatrix}.$$

It is clear that $|a_{nk}| \leq 1, n, k = 0, 1, 2, \ldots$ and $\sum_{k=0}^{\infty} a_{nk} = 1, n = 0, 1, 2, \ldots$

so that $\lim_{n \to \infty} \sum_{k=0}^{\infty} a_{nk} = 1$. Easy computation shows that $\lim_{n \to \infty} a_{nk}$ exists but $\neq 0$,

$k = 0, 1, 2, \ldots$. Using Theorem 4.1 $\bar{N} Y^{-1}$ is conservative but not regular. In view of Theorem 5.31, $c_Y \subseteq c_{\bar{N}}$ but $Y \not\subseteq \bar{N}$. Hence every bounded sequence which is Y-summable is also $\bar{N}$ summable but Y and $\bar{N}$ are not consistent for these sequences, proving that Brudno's theorem fails to hold in the ultrametric set-up. In this context, Somasundaram [18] also claims that Brudno's theorem fails to hold in the ultrametric set-up. However, his counterexample is incorrect and his proof is vitiated by an error involving the conclusion that if $|y'_n| \leq |y_n| + \lambda, n = 0, 1, 2, \ldots$ and $\{y_n\}$ converges, so does $\{y'_n\}$.

Definition 5.7   Given a sequence $\{x_k\}$, define the sequence $\{\bar{x}_k\}$ by

$$\bar{x}_0 = 0; \quad \bar{x}_k = x_{k-1}, \quad k \geq 1.$$

$A = (a_{nk})$ is said to be left translative if the A summability of $\{x_k\}$ to s implies the A summability of $\{\bar{x}_k\}$ to s. A is said to be right translative if the A summability of $\{\bar{x}_k\}$ to s implies the A summability of $\{x_k\}$ to s. If A is both left and right translative, A is said to be translative.

In [19], Natarajan obtained necessary and sufficient conditions for a regular $(\bar{N}, p_n)$ method to be left translative and right translative, which we shall record here without proofs.

Theorem 5.33 ([19], Theorem 3) $(\bar{N}, p_n)$ is left translative if and only if

$$\sup_n \left\{ \sup_{0 \leq k \leq n-2} \left| \frac{P_k}{P_n} \left(\frac{p_{k+1}}{p_k} - \frac{p_{k+2}}{p_{k+1}} \right) \right| \right\} < \infty. \tag{5.13}$$

Theorem 5.34 ([19], Theorem 4) $(\bar{N}, p_n)$ is right translative if and only if

$$\sup_n \left\{ \sup_{0 \leq k \leq n-1} \left| \frac{P_{k+1}}{P_n} \left(\frac{p_k}{p_{k+1}} - \frac{p_{k+1}}{p_{k+2}} \right) \right| \right\} < \infty. \tag{5.14}$$

Analogous to the classical situation (see [20], p. 17), we call the following theorem the "high indices theorem" for weighted means.

Theorem 5.35 ([21], Theorem 2) The $(\bar{N}, p_n)$ method is equivalent to convergence if and only if

$$\sup_n \left| \frac{P_n}{p_n} \right| < \infty. \tag{5.15}$$

We now prove a result which gives an equivalent formulation of summability by weighted mean methods. Incidentally, we note that this result includes the ultrametic analogue of a theorem proved by Môricz and Rhodes (see [22], Theorem MR, p. 188).

Theorem 5.36 ([23], Theorem 3) *Let* $(\overline{N}, p_n)$, $(\overline{N}, q_n)$ *be two regular weighted mean methods. Let* $\sum_{n=0}^{\infty} b_n$ *converge to* ℓ, *where*

$$b_n = q_n \sum_{k=n}^{\infty} \frac{x_k}{Q_k}, \quad n = 0, 1, 2, \ldots.$$

Then $\sum_{n=0}^{\infty} x_n$ *is* $(\overline{N}, p_n)$ *summable to* ℓ *if and only if*

$$\sup_{n,k} \left| \frac{p_k Q_{k+1}}{P_n q_{k+1}} \right| < \infty.$$

Proof Let $B_n = \sum_{k=0}^{n} b_k \to \ell$, $n \to \infty$. Now,

$$\frac{b_n}{q_n} - \frac{b_{n+1}}{q_{n+1}} = \sum_{k=n}^{\infty} \frac{x_k}{Q_k} - \sum_{k=n+1}^{\infty} \frac{x_k}{Q_k} = \frac{x_n}{Q_n},$$

so that

$$x_n = Q_n \left(\frac{b_n}{q_n} - \frac{b_{n+1}}{q_{n+1}} \right), \quad n = 0, 1, 2, \ldots.$$

Consequently,

$$\begin{aligned}
s_m = \sum_{k=0}^{m} x_k &= \sum_{k=0}^{m} Q_k \left(\frac{b_k}{q_k} - \frac{b_{k+1}}{q_{k+1}} \right) \\
&= \sum_{k=0}^{m} Q_k \frac{b_k}{q_k} - \sum_{k=1}^{m+1} Q_{k-1} \frac{b_k}{q_k} \\
&= Q_0 \frac{b_0}{q_0} + \sum_{k=1}^{m} (Q_k - Q_{k-1}) \frac{b_k}{q_k} - Q_m \frac{b_{m+1}}{q_{m+1}} \\
&= b_0 + \sum_{k=1}^{m} q_k \frac{b_k}{q_k} - Q_m \frac{b_{m+1}}{q_{m+1}}
\end{aligned}$$

$$= b_0 + \sum_{k=1}^{m} b_k - Q_m \frac{b_{m+1}}{q_{m+1}}$$

$$= \sum_{k=0}^{m} b_k - Q_m \frac{b_{m+1}}{q_{m+1}}$$

$$= B_m - Q_m \frac{b_{m+1}}{q_{m+1}}. \tag{5.16}$$

By hypothesis, $\displaystyle\sum_{k=0}^{\infty} \frac{x_k}{Q_k}$ converges so that

$$\frac{b_n}{q_n} = \sum_{k=n}^{\infty} \frac{x_k}{Q_k} \to 0, n \to \infty.$$

Now,

$$\frac{s_m}{Q_m} = \frac{B_m}{Q_m} - \frac{b_{m+1}}{q_{m+1}}, \quad \text{using (5.16)}.$$

Since $\{B_n\}$ converges, it is bounded so that $|B_n| \le M, n = 0, 1, 2, \dots$ for some $M > 0$. As $(\overline{N}, q_n)$ is regular, $|Q_n| \to \infty, n \to \infty$ in view of Theorem 5.23. Thus

$$\left| \frac{B_m}{Q_m} \right| \le \frac{M}{|Q_m|} \to 0, m \to \infty.$$

Consequently,

$$\frac{s_m}{Q_m} \to 0, m \to \infty.$$

For $n = 0, 1, 2, \dots,$

$$b_n = q_n \sum_{k=n}^{\infty} \frac{x_k}{Q_k} = q_n \lim_{m \to \infty} \sum_{k=n}^{m} \frac{s_k - s_{k-1}}{Q_k}$$

$$\text{(where } s_{-1} = 0)$$

$$= q_n \lim_{m \to \infty} \left\{ \sum_{k=n}^{m} \frac{s_k}{Q_k} - \sum_{k=n-1}^{m-1} \frac{s_k}{Q_{k+1}} \right\}$$

$$= q_n \lim_{m \to \infty} \left\{ \sum_{k=n}^{m-1} \frac{s_k}{Q_k} + \frac{s_m}{Q_m} - \sum_{k=n}^{m-1} \frac{s_k}{Q_{k+1}} - \frac{s_{n-1}}{Q_n} \right\}$$

$$= q_n \lim_{m \to \infty} \left\{ \sum_{k=n}^{m-1} \left(\frac{1}{Q_k} - \frac{1}{Q_{k+1}} \right) s_k + \frac{s_m}{Q_m} - \frac{s_{n-1}}{Q_n} \right\}$$

$$= -q_n \frac{s_{n-1}}{Q_n} + q_n \sum_{k=n}^{\infty} \left(\frac{1}{Q_k} - \frac{1}{Q_{k+1}} \right) s_k,$$

since $\lim_{m \to \infty} \frac{s_m}{Q_m} = 0$

$$= -q_n \frac{s_{n-1}}{Q_n} + q_n \sum_{k=n}^{\infty} u_k s_k, \text{ where } u_k = \frac{1}{Q_k} - \frac{1}{Q_{k+1}}. \quad (5.17)$$

Now,

$$B_n = \sum_{k=0}^{n-1} b_k + b_n$$

$$= \sum_{k=0}^{n-1} \frac{b_k}{q_k} q_k + b_n$$

$$= \sum_{k=0}^{n-1} q_k \left(\sum_{u=k}^{\infty} \frac{x_u}{Q_u} \right) + b_n$$

$$= q_0 \sum_{u=0}^{\infty} \frac{x_u}{Q_u} + q_1 \sum_{u=1}^{\infty} \frac{x_u}{Q_u} + q_2 \sum_{u=2}^{\infty} \frac{x_u}{Q_u} + \cdots + q_{n-1} \sum_{u=n-1}^{\infty} \frac{x_u}{Q_u} + b_n$$

$$= (q_0 + q_1 + \cdots + q_{n-1}) \sum_{u=n-1}^{\infty} \frac{x_u}{Q_u} + q_0 \sum_{u=0}^{n-2} \frac{x_u}{Q_u} + q_1 \sum_{u=1}^{n-2} \frac{x_u}{Q_u} + q_2 \sum_{u=2}^{n-2} \frac{x_u}{Q_u}$$

$$+ \cdots + q_{n-2} \frac{x_{n-2}}{Q_{n-2}} + b_n$$

$$= Q_{n-1} \sum_{u=n-1}^{\infty} \frac{x_u}{Q_u} + b_n + \frac{x_{n-2}}{Q_{n-2}} Q_{n-2} + \frac{x_{n-3}}{Q_{n-3}} Q_{n-3} + \cdots + \frac{x_0}{Q_0} Q_0$$

$$= Q_{n-1} \sum_{u=n-1}^{\infty} \frac{x_u}{Q_u} + b_n + \sum_{k=0}^{n-2} x_k$$

$$= s_{n-2} + b_n + Q_{n-1} \sum_{u=n-1}^{\infty} \frac{x_u}{Q_u}$$

$$= s_{n-2} + q_n \sum_{u=n}^{\infty} \frac{x_u}{Q_u} + Q_{n-1} \sum_{u=n-1}^{\infty} \frac{x_u}{Q_u}$$

$$= s_{n-2} + (Q_n - Q_{n-1}) \sum_{u=n}^{\infty} \frac{x_u}{Q_u} + Q_{n-1} \sum_{u=n-1}^{\infty} \frac{x_u}{Q_u}$$

$$= s_{n-2} + Q_n \sum_{u=n}^{\infty} \frac{x_u}{Q_u} + Q_{n-1} \frac{x_{n-1}}{Q_{n-1}}$$

$$= s_{n-1} + Q_n \sum_{u=n}^{\infty} \frac{x_u}{Q_u}$$

$$= s_{n-1} + Q_n \frac{b_n}{q_n}$$

$$= s_{n-1} + Q_n \left[-\frac{s_{n-1}}{Q_n} + \sum_{k=n}^{\infty} u_k s_k \right], \quad \text{using (5.17)}$$

$$= Q_n \sum_{k=n}^{\infty} u_k s_k,$$

so that

$$\frac{B_n}{Q_n} = \sum_{k=n}^{\infty} u_k s_k.$$

Consequently,

$$u_n s_n = \frac{B_n}{Q_n} - \frac{B_{n+1}}{Q_{n+1}}, \quad n = 0, 1, 2, \dots. \tag{5.18}$$

Let $\{T_n\}$ be the $(\overline{N}, p_n)$ transform of $\{s_k\}$ so that

$$T_n = \frac{1}{P_n} \sum_{k=0}^{n} p_k s_k$$

$$= \frac{1}{P_n} \sum_{k=0}^{n} p_k \frac{1}{u_k} \left\{ \frac{B_k}{Q_k} - \frac{B_{k+1}}{Q_{k+1}} \right\}, \quad \text{using (5.18)}$$

$$= \frac{1}{P_n} \left[\frac{p_0}{u_0} \frac{B_0}{Q_0} + \sum_{k=1}^{n} \left\{ \frac{p_k}{u_k} - \frac{p_{k-1}}{u_{k-1}} \right\} \frac{B_k}{Q_k} - \frac{p_n}{u_n} \frac{B_{n+1}}{Q_{n+1}} \right]$$

$$= \sum_{k=0}^{\infty} a_{nk} B_k,$$

where

$$a_{nk} = \begin{cases} \frac{1}{P_n} \frac{p_0}{u_0 Q_0}, & k = 0; \\ \frac{1}{P_n} \left\{ \frac{p_k}{u_k} - \frac{p_{k-1}}{u_{k-1}} \right\} \frac{1}{Q_k}, & 1 \leq k \leq n; \\ -\frac{1}{P_n} \frac{p_n}{u_n Q_{n+1}}, & k = n + 1; \\ 0, & k \geq n + 2. \end{cases}$$

It is clear that $\lim\limits_{n \to \infty} a_{nk} = 0$, $k = 0, 1, 2, \dots$. Also,

$$\sum_{k=0}^{\infty} a_{nk} = \sum_{k=0}^{n+1} a_{nk}$$

$$= \frac{1}{P_n} \left[\frac{p_0}{u_0 Q_0} + \sum_{k=1}^{n} \left\{ \frac{p_k}{u_k} - \frac{p_{k-1}}{u_{k-1}} \right\} \frac{1}{Q_k} - \frac{p_n}{u_n Q_{n+1}} \right]$$

$$= \frac{1}{P_n} \left[\frac{p_0}{u_0 Q_0} + \left(\frac{p_1}{u_1} - \frac{p_0}{u_0} \right) \frac{1}{Q_1} + \left(\frac{p_2}{u_2} - \frac{p_1}{u_1} \right) \frac{1}{Q_2} \right.$$

$$\left. + \cdots + \left(\frac{p_n}{u_n} - \frac{p_{n-1}}{u_{n-1}} \right) \frac{1}{Q_n} - \frac{p_n}{u_n Q_{n+1}} \right]$$

$$= \frac{1}{P_n} \left[\frac{p_0}{u_0} \left(\frac{1}{Q_0} - \frac{1}{Q_1} \right) + \frac{p_1}{u_1} \left(\frac{1}{Q_1} - \frac{1}{Q_2} \right) + \frac{p_2}{u_2} \left(\frac{1}{Q_2} - \frac{1}{Q_3} \right) \right.$$

$$\left. + \cdots + \frac{p_n}{u_n} \left(\frac{1}{Q_n} - \frac{1}{Q_{n+1}} \right) \right]$$

$$= \frac{1}{P_n} \left[\frac{p_0}{u_0} u_0 + \frac{p_1}{u_1} u_1 + \cdots + \frac{p_n}{u_n} u_n \right]$$

$$= \frac{1}{P_n} P_n$$

$$= 1, \quad n = 0, 1, 2, \ldots,$$

so that $\lim_{n \to \infty} \sum_{k=0}^{\infty} a_{nk} = 1$. By hypothesis, $B_k \to \ell, k \to \infty$. Using Theorem 4.1,

$T_n \to \ell, n \to \infty$, i.e. $\sum_{n=0}^{\infty} x_n$ is $(\overline{N}, p_n)$ summable to ℓ if and only if

$$\sup_n \frac{1}{|P_n|} \left[\frac{1}{|Q_0|} \left| \frac{p_0}{u_0} \right|, \max_{1 \le k \le n} \frac{1}{|Q_k|} \left\{ \left| \frac{p_k}{u_k} - \frac{p_{k-1}}{u_{k-1}} \right| \right\}, \frac{1}{|Q_{n+1}|} \left| \frac{p_n}{u_n} \right| \right] < \infty,$$

i.e. if and only if

$$\sup_n \frac{1}{|P_n|} \left[\max_{1 \le k \le n} \left| \frac{p_k Q_{k+1}}{q_{k+1}} - \frac{p_{k-1} Q_{k-1}}{q_k} \right|, \left| \frac{p_n Q_n}{q_{n+1}} \right| \right] < \infty,$$

which is equivalent to

$$\sup_n \frac{1}{|P_n|} \left[\max_{1 \le k \le n} \left| \frac{p_k Q_{k+1}}{q_{k+1}} - \frac{p_{k-1} Q_{k-1}}{q_k} \right| \right] < \infty$$

(here we use the fact that $|P_n| \le |P_{n+1}|$, $|Q_n| \le |Q_{n+1}|$, $n = 0, 1, 2, \ldots$). This completes the proof of the theorem. □

Following Môricz and Rhoades [22], we now prove another interesting result on weighted mean methods (see [24]).

Theorem 5.37 *Let* $(\overline{N}, p_n)$, $(\overline{N}, q_n)$ *be regular weighted mean methods and*

$$P_n = O(p_n Q_n), \quad n \to \infty, \tag{5.19}$$

in the sense that $\left|\frac{P_n}{p_n Q_n}\right| \leq M$ *for some* $M > 0$, $n = 0, 1, 2, \ldots$. *Let* $\sum_{n=0}^{\infty} x_n$ *be* $(\overline{N}, p_n)$ *summable to* ℓ. *Then* $\sum_{n=0}^{\infty} b_n$ *converges to* ℓ *if and only if*

$$\sup_{n} \left[|Q_n| \sup_{k \geq n} \left| \frac{P_k}{Q_{k+1}} \left(\frac{q_{k+1}}{p_k Q_k} - \frac{q_{k+2}}{p_{k+1} Q_{k+2}} \right) \right| \right] < \infty,$$

where $b_n = q_n \sum_{k=n}^{\infty} \frac{x_k}{Q_k}$, $n = 0, 1, 2, \ldots$.

Proof Let $t_n = \frac{p_0 s_0 + p_1 s_1 + \cdots + p_n s_n}{P_n}$, $n = 0, 1, 2, \ldots$, where $s_n = \sum_{k=0}^{n} x_k$, $n = 0, 1, 2, \ldots$. Then $s_0 = t_0$ and $s_n = \frac{1}{p_n}(P_n t_n - P_{n-1} t_{n-1})$, $n = 0, 1, 2, \ldots$. Let $\sum_{n=0}^{\infty} x_n$ be $(\overline{N}, p_n)$ summable to ℓ so that $\lim_{n \to \infty} t_n = \ell$. Now,

$$\begin{aligned}
\frac{s_n}{Q_n} &= \frac{1}{p_n Q_n}(P_n t_n - P_{n-1} t_{n-1}) \\
&= \frac{1}{p_n Q_n}\left[P_n(t_n - \ell) - P_{n-1}(t_{n-1} - \ell) + \ell(P_n - P_{n-1}) \right] \\
&= \frac{1}{p_n Q_n}\left[P_n(t_n - \ell) - P_{n-1}(t_{n-1} - \ell) + \ell p_n \right] \\
&= \frac{P_n}{p_n Q_n}(t_n - \ell) - \frac{P_{n-1}}{p_n Q_n}(t_{n-1} - \ell) + \frac{\ell}{Q_n}
\end{aligned}$$

so that

$$\left| \frac{s_n}{Q_n} \right| \leq \max\left[M\left(|t_n - \ell|, |t_{n-1} - \ell| \right), \frac{|\ell|}{|Q_n|} \right],$$

since $|P_{n-1}| \leq |P_n|$

$\to 0, n \to \infty$, since $\lim_{n \to \infty} t_n = \ell$ and $\lim_{n \to \infty} |Q_n| = \infty$, $(\overline{N}, p_n)$ being regular, using Theorem 5.23.

As already worked out in the proof of Theorem 5.36,

$$b_n = -\frac{q_n s_{n-1}}{Q_n} + q_n \sum_{k=n}^{\infty} c_k s_k,$$

where

$$c_k = \frac{1}{Q_k} - \frac{1}{Q_{k+1}}, \quad k = 0, 1, 2, \ldots.$$

Now,

$$B_n = s_{n-1} + Q_n \left(-\frac{s_{n-1}}{Q_n} + \sum_{k=n}^{\infty} c_k s_k \right) \text{(see proof of Theorem 5.36)}$$

$$= s_{n-1} - s_{n-1} + Q_n \sum_{k=n}^{\infty} c_k s_k$$

$$= Q_n \sum_{k=n}^{\infty} c_k s_k$$

$$= Q_n \lim_{m \to \infty} \sum_{k=n}^{m} c_k s_k$$

$$= Q_n \lim_{m \to \infty} \sum_{k=n}^{m} c_k \frac{1}{p_k} \{P_k t_k - P_{k-1} t_{k-1}\}$$

$$= Q_n \lim_{m \to \infty} \left[\frac{c_m P_m t_m}{p_m} - \frac{c_n P_{n-1} t_{n-1}}{p_n} + \sum_{k=n}^{m-1} P_k t_k \left(\frac{c_k}{p_k} - \frac{c_{k+1}}{p_{k+1}} \right) \right]. \quad (5.20)$$

Let

$$A_1 = \left\{ \{x_k\} : \sum_{k=0}^{\infty} x_k \text{ is } (\overline{N}, p_n) \text{ summable} \right\};$$

$$A_2 = \left\{ \{x_k\} : \sum_{k=0}^{\infty} b_k \text{ converges} \right\}.$$

We note that A_1, A_2 are ultrametric BK spaces with respect to the norms defined by

$$\|x\|_{A_1} = \sup_{n \geq 0} |t_n|, \quad x = \{x_k\} \in A_1;$$

and
$$\|x\|_{A_2} = \sup_{n \geq 0} |B_n|, \quad x = \{x_k\} \in A_2,$$

respectively. In view of Banach–Steinhaus theorem (see [25]),

$$\|x\|_{A_2} \leq L \|x\|_{A_1}, \quad \text{for some } L > 0. \tag{5.21}$$

For every fixed $k = 0, 1, 2, \ldots$, define the sequence $x = \{x_n\}$, where

$$x_n = \begin{cases} 1, & \text{if } n = k; \\ -1, & \text{if } n = k+1; \\ 0, & \text{otherwise.} \end{cases}$$

For this sequence x,

$$\|x\|_{A_1} = \left| \frac{p_k}{P_k} \right| \quad \text{and} \quad \|x\|_{A_2} = |Q_k c_k|.$$

Using (5.21), we have, for $k = 0, 1, 2, \ldots$, $|Q_k c_k| \leq L \frac{|p_k|}{|P_k|}$, so that

$$\left| \frac{c_k P_k}{p_k} \right| \leq \frac{L}{|Q_k|} \to 0, k \to \infty, \quad \text{since} \quad \lim_{k \to \infty} |Q_k| = \infty,$$

in view of Theorem 5.23. Consequently,

$$\lim_{k \to \infty} \frac{c_k P_k}{p_k} = 0. \tag{5.22}$$

Using (5.22) in (5.20), we have,

$$B_n = -\frac{c_n P_{n-1} t_{n-1}}{p_n} Q_n + Q_n \sum_{k=n}^{\infty} P_k t_k \left(\frac{c_k}{p_k} - \frac{c_{k+1}}{p_{k+1}} \right)$$

$$= \sum_{k=0}^{\infty} a_{nk} t_k,$$

where the infinite matrix (a_{nk}) is defined by

$$a_{nk} = \begin{cases} 0, & 0 \leq k < n-1; \\ -\frac{Q_n c_n P_{n-1}}{p_n}, & k = n-1; \\ Q_n P_k \left(\frac{c_k}{p_k} - \frac{c_{k+1}}{p_{k+1}} \right), & k \geq n. \end{cases}$$

We first note that $\lim_{n \to \infty} a_{nk} = 0, k = 0, 1, 2, \ldots$ and $\sum_{k=0}^{\infty} a_{nk} = 1, n = 0, 1, 2, \ldots$

so that $\lim_{n \to \infty} \sum_{k=0}^{\infty} a_{nk} = 1$. Thus, appealing to Theorem 4.1, $\sum_{n=0}^{\infty} b_n$ converges to ℓ if and only if

$$\sup_{n \geq 0} |Q_n| \left[\max \left\{ \left| \frac{c_n P_{n-1}}{p_n} \right|, \sup_{k \geq n} \left| P_k \left(\frac{c_k}{p_k} - \frac{c_{k+1}}{p_{k+1}} \right) \right| \right\} \right] < \infty. \tag{5.23}$$

However,

$$\left| \frac{Q_n P_{n-1} c_n}{p_n} \right| \leq \left| \frac{Q_n P_n c_n}{p_n} \right|, \quad \text{since } |P_{n-1}| \leq |P_n|$$

$$= |Q_n| \left| P_n \sum_{k=n}^{\infty} \left(\frac{c_k}{p_k} - \frac{c_{k+1}}{p_{k+1}} \right) \right|, \quad \text{using (5.22)}$$

$$\leq |Q_n| \left| \sum_{k=n}^{\infty} P_k \left(\frac{c_k}{p_k} - \frac{c_{k+1}}{p_{k+1}} \right) \right|, \quad \text{since } |P_n| \leq |P_k|, k \geq n$$

$$\leq |Q_n| \sup_{k \geq n} \left| P_k \left(\frac{c_k}{p_k} - \frac{c_{k+1}}{p_{k+1}} \right) \right|. \tag{5.24}$$

Using (5.24), it is now clear that (5.23) is equivalent to

$$\sup_{n \geq 0} |Q_n| \left[\sup_{k \geq n} \left| P_k \left(\frac{c_k}{p_k} - \frac{c_{k+1}}{p_{k+1}} \right) \right| \right] < \infty.$$

Now,

$$\frac{c_k}{p_k} - \frac{c_{k+1}}{p_{k+1}} = \frac{1}{p_k} \left(\frac{1}{Q_k} - \frac{1}{Q_{k+1}} \right) - \frac{1}{p_{k+1}} \left(\frac{1}{Q_{k+1}} - \frac{1}{Q_{k+2}} \right)$$

$$= \frac{1}{p_k} \left(\frac{Q_{k+1} - Q_k}{Q_k Q_{k+1}} \right) - \frac{1}{p_{k+1}} \left(\frac{Q_{k+2} - Q_{k+1}}{Q_{k+1} Q_{k+2}} \right)$$

$$= \frac{q_{k+1}}{p_k Q_k Q_{k+1}} - \frac{q_{k+2}}{p_{k+1} Q_{k+1} Q_{k+2}}.$$

Thus $\sum_{n=0}^{\infty} b_n$ converges to ℓ if and only if

$$\sup_{n \geq 0} |Q_n| \left[\sup_{k \geq n} \left| P_k \left(\frac{q_{k+1}}{p_k Q_k Q_{k+1}} - \frac{q_{k+2}}{p_{k+1} Q_{k+1} Q_{k+2}} \right) \right| \right] < \infty.$$

$$\text{i.e.} \quad \sup_{n \geq 0} |Q_n| \left[\sup_{k \geq n} \left| \frac{P_k}{Q_{k+1}} \left(\frac{q_{k+1}}{P_k Q_k} - \frac{q_{k+2}}{P_{k+1} Q_{k+2}} \right) \right| \right] < \infty,$$

completing the proof of the theorem.   □

References

1. Srinivasan, V.K.: On certain summation processes in the p-adic field. Indag. Math. **27**, 319–325 (1965)
2. Natarajan, P.N.: On Nörlund method of summability in non-archimedean fields. J. Anal. **2**, 97–102 (1994)
3. Hardy, G.H.: Divergent Series. Oxford (1949)
4. Natarajan, P.N.: Some theorems on Cauchy multiplication for Nörlund means in non-archimedean fields. J. Orissa Math. Soc. **12–15**, 33–37 (1993–1996)
5. Natarajan, P.N.: A multiplication theorem for Nörlund means in non-archimedean fields. J. Anal. **5**, 59–63 (1997)
6. Natarajan, P.N.: Multiplication theorems for Nörlund means in non-archimedean fields II. J. Anal. **6**, 49–53 (1998)
7. Escassut, A.: Analytic Elements in p-adic Analysis. World Scientific (1995)
8. Natarajan, P.N., Srinivasan, V.: On the convergence field of Nörlund means in non-archimedean fields. J. Anal. **13**, 25–30 (2005)
9. Wilansky, A.: Functional Analysis. Blaisdell (1964)
10. Natarajan, P.N., Sakthivel, S.: Convergence fields of Nörlund means in non-archimedean fields. J. Approx. Theor. Appl. **2**, 133–147 (2006)
11. Wilansky, A.: Summability Through Functional Analysis. North Holland, Amsterdam (1984)
12. Natarajan, P.N.: On conservative non-regular Nörlund and weighted means in non-archimedean fields. Ramanujan J. **4**, 429–433 (2000)
13. Natarajan, P.N.: A note on Schur matrices in non-archimedean fields. J. Ramanujan Math. Soc. **10**, 157–162 (1995)
14. Natarajan, P.N.: Some properties of regular Nörlund methods in non-archimedean fields. Indian J. Math. **53**, 287–299 (2011)
15. Natarajan, P.N.: Weighted means in non-archimedean fields. Ann. Math. Blaise Pascal **2**, 191–200 (1995)
16. Natarajan, P.N.: Failure of two classical summability theorems in non-archimedean analysis. J. Anal. **7**, 1–5 (1999)
17. Cooke, R.G.: Infinite Matrices and Sequence Spaces. Macmillan (1950)
18. Somasundaram, D.: On a theorem of Brudno over non-archimedean fields. Bull. Austral. Math. Soc. **23**, 191–194 (1981)
19. Natarajan, P.N.: Translativity of weighted means in non-archimedean fields. Indian J. Math. **47**, 123–130 (2005)
20. Peyerimhoff, A.: Lectures on summability. Lecture Notes in Mathematics, vol. 107. Springer (1969)
21. Natarajan, P.N.: More about $(\bar{N}, p_n)$ methods in non-archimedean fields. Indian J. Math. **46**, 87–100 (2004)
22. Móricz, F., Rhoades, B.E.: An equivalent reformulation of summability by weighted mean methods, revisited. Linear Algebra Appl. **349**, 187–192 (2002)
23. Natarajan, P.N.: A theorem on weighted means in non-archimedean fields, p-adic Numbers. Ultrametric Anal. Appl. **2**, 363–367 (2010)
24. Natarajan, P.N.: Another theorem on weighted means in non-archimedean fields, p-adic Numbers. Ultrametric Anal. Appl. **3**, 81–85 (2011)
25. Monna, A.F.: Sur le théorème de Banach-Steinhaus. Indag. Math. **25**, 121–131 (1963)

Chapter 6
The Euler and The Taylor Methods

Abstract In this chapter, we introduce the Euler and the Taylor methods and present a detailed study of their properties.

Keywords The Euler method · The Taylor method

We now define the Euler and Taylor methods of summability in ultrametric analysis and record some of their properties (see [1]).

Definition 6.1 Let $r \in K$ be such that $|1 - r| < 1$. The Euler method of order r or the (E, r) method is given by the matrix $(e_{nk}^{(r)})$, $n, k = 0, 1, 2, \ldots$ which is defined as follows:

If $r \neq 1$,

$$e_{nk}^{(r)} = \begin{cases} c_n^k r^k (1 - r)^{n-k}, & k \leq n; \\ 0, & k > n, \end{cases}$$

where $c_n^k = \frac{n!}{k!(n-k)!}$, $k \leq n$;

If $r = 1$,

$$e_{nk}^{(r)} = \begin{cases} 1, & k = n; \\ 0, & k \neq n. \end{cases}$$

$(e_{nk}^{(r)})$ is called the (E, r) matrix.

Sometimes, for convenience, we also use the notation $^n c_k$ for c_n^k.

Theorem 6.1 ([1], Theorem 1.2) *The (E, r) method is regular.*

Proof Since $|1 - r| < 1$, $|r| = |(r - 1) + 1| = 1$, since the valuation is ultrametric. If $e_{nk}^{(r)} \neq 0$,

© Springer India 2015
P.N. Natarajan, *An Introduction to Ultrametric Summability Theory*,
Forum for Interdisciplinary Mathematics 2, DOI 10.1007/978-81-322-2559-1_6

$$|e_{nk}^{(r)}| = |c_n^k r^k (1 - r)^{n-k}|$$
$$\leq |r|^k |1 - r|^{n-k}, \quad \text{since } |c_n^k| \leq 1$$
$$= |1 - r|^{n-k}, \quad \text{since } |r| = 1$$
$$< 1,$$

and so

$$\sup_{n,k} |e_{nk}^{(r)}| < 1.$$

Now, for $k = 0, 1, 2, \ldots,$

$$\lim_{n \to \infty} |e_{nk}^{(r)}| = \lim_{n \to \infty} |c_n^k r^k (1 - r)^{n-k}|$$
$$\leq \lim_{n \to \infty} |1 - r|^{n-k}$$
$$= 0, \quad \text{since } |1 - r| < 1.$$

Consequently,

$$\lim_{n \to \infty} e_{nk}^{(r)} = 0, \quad k = 0, 1, 2, \ldots.$$

Also, using Binomial theorem (see [2], p. 51), we have,

$$\sum_{k=0}^{\infty} e_{nk}^{(r)} = \sum_{k=0}^{n} c_n^k r^k (1 - r)^{n-k}$$
$$= (1 - r + r)^n$$
$$= 1, \quad n = 0, 1, 2, \ldots, \quad \text{since } |n| \leq 1,$$

so that

$$\lim_{n \to \infty} \sum_{k=0}^{\infty} e_{nk}^{(r)} = 1.$$

Thus the (E, r) method is regular. $\square$

Theorem 6.2 ([1], Theorem 1.3) $(e_{nk}^{(r)})(e_{nk}^{(s)})$ is the (E, rs) matrix.

Proof Let $(a_{nj}) = (e_{nk}^{(r)})(e_{nk}^{(s)})$. It is clear that $a_{nj} = 0, j > n$. If r or $s = 1$, the result follows since $(E, 1)$ is usual convergence. Let now, $r, s \neq 1$ such that $|1 - r|, |1 - s| < 1$.

Let $j \leq n$. Now,

$$a_{nj} = \sum_{k=j}^{n} e_{nk}^{(r)} e_{kj}^{(s)},$$

noting that $e_{kj}^{(s)} = 0, k < j$. If $j = n$,

$$a_{nn} = e_{nn}^{(r)} e_{nn}^{(s)} = r^n s^n = (rs)^n;$$

If $j < n$,

$$a_{nj} = \sum_{k=j}^{n} e_{nk}^{(r)} e_{kj}^{(s)}$$

$$= \sum_{k=j}^{n} c_n^k r^k (1-r)^{n-k} c_k^j s^j (1-s)^{k-j}$$

$$= s^j (1-r)^n \sum_{k=j}^{n} c_n^k c_k^j r^k \frac{(1-s)^{k-j}}{(1-r)^k}$$

$$= c_n^j s^j r^j (1-r)^{n-j} \sum_{k=j}^{n} c_{(n-j)}^{(k-j)} r^{k-j} \frac{(1-s)^{k-j}}{(1-r)^{k-j}}$$

$$= c_n^j (rs)^j (1-r)^{n-j} \sum_{k=j}^{n} c_{(n-j)}^{(k-j)} \left[\frac{r(1-s)}{1-r} \right]^{k-j}$$

$$= c_n^j (rs)^j (1-r)^{n-j} \left[1 + \frac{r(1-s)}{1-r} \right]^{n-j}$$

$$= c_n^j (rs)^j (1-rs)^{n-j},$$

completing the proof of the theorem.   □

Remark 6.1 According to Definition 6.1, r belongs to the disc $D = \{x \in K / |x - 1| < 1\}$. This disc is a multiplicative group, noting that $|1-r|, |1-s| < 1$ imply $|1 - rs| < 1$. Now the above theorem says: The subjective mapping from D onto the set of Euler matrices, which associates (E, r) to each $r \in D$, is a group homomorphism.

Corollary 6.1 *The (E, r) matrix is invertible and its inverse is the $\left(E, \frac{1}{r} \right)$ matrix.*

Theorem 6.3 ([1], Theorem 1.5) *If $|r - s| < |r|$, then $(E, r) \subseteq (E, s)$.*

Proof Let $\{z_k\}$ be (E, r) summable to y, i.e. $\{t_n\}$ converges to y, where

$$t_n = \sum_{k=0}^{n} e_{nk}^{(r)} z_k, \quad n = 0, 1, 2, \ldots.$$

Now,

$$\sum_{n=0}^{j} e_{jn}^{(\frac{1}{r})} t_n = \sum_{n=0}^{j} e_{jn}^{(\frac{1}{r})} \left(\sum_{k=0}^{n} e_{nk}^{(r)} z_k \right)$$

$$= \sum_{k=0}^{j} \left(\sum_{n=k}^{j} e_{jn}^{(\frac{1}{r})} e_{nk}^{(r)} \right) z_k$$

$$= z_j,$$

in view of corollary of Theorem 6.2. Again,

$$\sigma_k = \sum_{j=0}^{k} e_{kj}^{(s)} z_k$$

$$= \sum_{j=0}^{k} e_{kj}^{(s)} \left(\sum_{n=0}^{j} e_{jn}^{(\frac{1}{r})} t_n \right)$$

$$= \sum_{n=0}^{k} \left(\sum_{j=n}^{k} e_{kj}^{(s)} e_{jn}^{(\frac{1}{r})} \right) t_n$$

$$= \sum_{n=0}^{k} e_{kn}^{(\frac{s}{r})} t_n,$$

using Theorem 6.2. Since $|r - s| < |r|$, $\left| \frac{s}{r} - 1 \right| < 1$, the method $\left(E, \frac{s}{r} \right)$ is regular. Since $\{t_n\}$ converges to y and $\left(E, \frac{s}{r} \right)$ is regular, $\{\sigma_k\}$ converges to y, i.e. $\{z_k\}$ is (E, s) summable to y. Thus $(E, r) \subseteq (E, s)$, completing the proof of the theorem. □

We now prove a pair of theorems on the Cauchy multiplication of Euler summable sequences and series [3].

Theorem 6.4 If $x_k = o(1)$, $k \to \infty$ and $\{y_k\}$ is (E, r) summable to σ, then $\{z_k\}$ is (E, r) summable to

$$\sigma \left[x_0 + \sum_{k=1}^{\infty} x_k r^{k-1} \right],$$

where $z_n = \sum_{k=0}^{n} x_k y_{n-k}$, $n = 0, 1, 2, \ldots$.

Proof Let $\{\sigma_n\}$ be the (E, r) transform of $\{y_k\}$. Then,

$$\sigma_n = \sum_{k=0}^{n} {}^n c_k r^k (1 - r)^{n-k} y_k, \quad n = 0, 1, 2, \ldots. \tag{6.1}$$

By hypothesis, $\lim\limits_{n\to\infty}\sigma_n=\sigma$. Let $\{\tau_n\}$ be the (E,r) transform of $\{z_k\}$ so that

$$
\begin{aligned}
\tau_n &= \sum_{k=0}^{n} {}^nc_k r^k (1-r)^{n-k} z_k \\
&= (1-r)^n z_0 + {}^nc_1 r(1-r)^{n-1} z_1 + {}^nc_2 r^2 (1-r)^{n-2} z_2 + \cdots + r^n z_n \\
&= (1-r)^n (x_0 y_0) + {}^nc_1 r(1-r)^{n-1}(x_0 y_1 + x_1 y_0) \\
&\quad + {}^nc_2 r^2 (1-r)^{n-2}(x_0 y_2 + x_1 y_1 + x_2 y_0) \\
&\quad + \cdots + r^n (x_0 y_n + x_1 y_{n-1} + \cdots + x_n y_0)) \\
&= x_0 [(1-r)^n y_0 + {}^nc_1 r(1-r)^{n-1} y_1 + {}^nc_2 r^2 (1-r)^{n-2} y_2 + \cdots + r^n y_n] \\
&\quad + x_1 [{}^nc_1 r(1-r)^{n-1} y_0 + {}^nc_2 r^2 (1-r)^{n-2} y_1 + \cdots + r^n y_{n-1}] \\
&\quad + x_2 [{}^nc_2 r^2 (1-r)^{n-2} y_0 + {}^nc_3 r^3 (1-r)^{n-3} y_1 + \cdots + r^n y_{n-2}] \\
&\quad + \cdots + x_n r^n y_0 \\
&= x_0 \left[\sum_{k=0}^{n} {}^nc_k r^k (1-r)^{n-k} y_k \right] + x_1 \left[\sum_{k=1}^{n} {}^nc_k r^k (1-r)^{n-k} y_{k-1} \right] \\
&\quad + x_2 \left[\sum_{k=2}^{n} {}^nc_k r^k (1-r)^{n-k} y_{k-2} \right] + \cdots + x_n r^n y_0 \\
&= x_0 \sigma_n + x_1 \left[\sum_{k=1}^{n} {}^nc_k r^k (1-r)^{n-k} y_{k-1} \right] \\
&\quad + x_2 \left[\sum_{k=2}^{n} {}^nc_k r^k (1-r)^{n-k} y_{k-2} \right] + \cdots + x_n r^n \sigma_0. \qquad (6.2)
\end{aligned}
$$

Now,

$$
\begin{aligned}
\sum_{k=1}^{n} {}^nc_k r^k (1-r)^{n-k} y_{k-1} &= \sum_{j=0}^{n-1} {}^nc_{j+1} r^{j+1} (1-r)^{n-j-1} y_j \\
&= \sum_{j=0}^{n-1} \left[{}^nc_{j+1} r^{j+1} (1-r)^{n-j-1} \right. \\
&\qquad \left. \times \left\{ \sum_{k=0}^{j} {}^jc_k \left(\frac{1}{r}\right)^k \left(1-\frac{1}{r}\right)^{j-k} \sigma_k \right\} \right],
\end{aligned}
$$

using Corollary 6.1 and (6.1)

$$
= \sum_{k=0}^{n-1} \left[r(1-r)^{n-k-1} \sigma_k \left\{ \sum_{j=k}^{n-1} (-1)^{j-k}\, {}^nc_{j+1}\, {}^jc_k \right\} \right],
$$

interchanging the order of summation. $\qquad (6.3)$

Using the identity

$$\sum_{k=0}^{n-1} \left(\sum_{j=k}^{n-1} (-1)^{j-k} \, {}^n c_{j+1} \, {}^j c_k \right) z^k = \sum_{k=0}^{n-1} z^k,$$

we have,

$$\sum_{j=k}^{n-1} (-1)^{j-k} \, {}^n c_{j+1} \, {}^j c_k = 1, \quad 0 \le k \le n-1. \tag{6.4}$$

Thus, using (6.3), (6.4), we have,

$$\sum_{k=1}^{n} {}^n c_k r^k (1-r)^{n-k} y_{k-1} = \sum_{k=0}^{n-1} r(1-r)^{n-k-1} \sigma_k. \tag{6.5}$$

Using (6.5) and similar results, (6.2) can now be written as

$$\tau_n = x_0 \sigma_n + x_1 \left(\sum_{k=0}^{n-1} r(1-r)^{n-k-1} \sigma_k \right)$$

$$+ x_2 \left(\sum_{k=0}^{n-2} r^2 (1-r)^{n-k-2} \sigma_k \right) + \cdots + x_n r^n \sigma_0$$

$$= x_0 (\sigma_n - \sigma) + x_1 \left\{ \sum_{k=0}^{n-1} r(1-r)^{n-k-1} (\sigma_k - \sigma) \right\}$$

$$+ x_2 \left\{ \sum_{k=0}^{n-2} r^2 (1-r)^{n-k-2} (\sigma_k - \sigma) \right\} + \cdots + x_n r^n (\sigma_0 - \sigma)$$

$$+ \sigma \left[x_0 + x_1 \left\{ \sum_{k=0}^{n-1} r(1-r)^{n-k-1} \right\} \right.$$

$$+ x_2 \left\{ \sum_{k=0}^{n-2} r^2 (1-r)^{n-k-2} \right\} + \cdots + x_n r^n \right]$$

$$= x_0 (\sigma_n - \sigma) + x_1 \left\{ \sum_{k=0}^{n-1} r(1-r)^{n-k-1} (\sigma_k - \sigma) \right\}$$

$$+ x_2 \left\{ \sum_{k=0}^{n-2} r^2 (1-r)^{n-k-2} (\sigma_k - \sigma) \right\} + \cdots + x_n r^n (\sigma_0 - \sigma)$$

$$+ \sigma \left[x_0 + x_1 r \left\{ \frac{1-(1-r)^n}{1-(1-r)} \right\} + x_2 r^2 \left\{ \frac{1-(1-r)^{n-1}}{1-(1-r)} \right\} + \cdots + x_n r^n \right]$$

$$= x_0(\sigma_n - \sigma) + x_1 \left\{ \sum_{k=0}^{n-1} r(1-r)^{n-k-1}(\sigma_k - \sigma) \right\}$$

$$+ x_2 \left\{ \sum_{k=0}^{n-2} r^2(1-r)^{n-k-2}(\sigma_k - \sigma) \right\} + \cdots + x_n r^n(\sigma_0 - \sigma)$$

$$+ \sigma \left[x_0 + x_1\{1 - (1-r)^n\} + x_2 r\{1 - (1-r)^{n-1}\} \right.$$

$$\left. + \cdots + x_n r^{n-1}\{1 - (1-r)\} \right]$$

$$= x_0(\sigma_n - \sigma) + x_1 r \left\{ \sum_{k=0}^{n-1}(1-r)^{n-k-1}(\sigma_k - \sigma) \right\}$$

$$+ x_2 r^2 \left\{ \sum_{k=0}^{n-2}(1-r)^{n-k-2}(\sigma_k - \sigma) \right\} + \cdots + x_n r^n(\sigma_0 - \sigma)$$

$$+ \sigma \left[(x_0 + x_1 + x_2 r + \cdots + x_n r^{n-1}) \right.$$

$$\left. - \{x_1(1-r)^n + x_2 r(1-r)^{n-1} + \cdots + x_n r^{n-1}(1-r)\} \right]. \tag{6.6}$$

Now,

$$|x_n r^{n-1}| = |x_n|, \quad \text{since } |r| = 1$$
$$\to 0, n \to \infty$$

and

$$|(1-r)^n| = |1-r|^n \to 0, n \to \infty, \quad \text{since } |1-r| < 1.$$

Thus $\{x_n r^{n-1}\}$ and $\{(1-r)^n\}$ are null sequences.

Note that the sequence

$$\{x_1(1-r)^n + x_2 r(1-r)^{n-1} + \cdots + x_n r^{n-1}(1-r)\}$$

is the Cauchy product of $\{x_n r^{n-1}\}$ and $\{(1-r)^n\}$.

In view of Theorem 4.16

$$\lim_{n \to \infty} \{x_1(1-r)^n + x_2 r(1-r)^{n-1} + \cdots + x_n r^{n-1}(1-r)\} = 0.$$

Let $\alpha_n = x_n r^n$. Note that $\{\alpha_n\}$ is a null sequence since $|r| = 1$ and $x_n \to 0, n \to \infty$. Let

$$\beta_n = \sum_{k=0}^{n-1}(1-r)^{n-k-1}(\sigma_k - \sigma).$$

Now, $\{\beta_n\}$ is the Cauchy product of the null sequences $\{(1 - r)^n\}$ and $\{\sigma_n - \sigma\}$. In view of Theorem 4.16,

$$\beta_n \to 0, n \to \infty.$$

We now note that

$$\lim_{n \to \infty} \left[x_1 r \left\{ \sum_{k=0}^{n-1} (1 - r)^{n-k-1} (\sigma_k - \sigma) \right\} \right.$$
$$\left. + x_2 r^2 \left\{ \sum_{k=0}^{n-2} (1 - r)^{n-k-2} (\sigma_k - \sigma) \right\} + \cdots + x_n r^n (\sigma_0 - \sigma) \right] = 0,$$

since $\lim_{n \to \infty} \alpha_n = 0 = \lim_{n \to \infty} \beta_n$, again appealing to Theorem 4.16. Thus, taking limit as $n \to \infty$ in (6.6), we have,

$$\lim_{n \to \infty} \tau_n = \sigma \left[x_0 + \sum_{k=1}^{\infty} x_k r^{k-1} \right],$$

noting that the series on the right converges since $|r| = 1$ and $x_k \to 0, k \to \infty$. In other words, $\{z_k\}$ is (E, r) summable to

$$\sigma \left[x_0 + \sum_{k=1}^{\infty} x_k r^{k-1} \right],$$

completing the proof of the theorem.   □

The following result can be proved in a similar fashion:

Theorem 6.5 *If* $\sum_{k=0}^{\infty} x_k$ *converges and* $\sum_{k=0}^{\infty} y_k$ *is* (E, r) *summable to* σ, *then* $\sum_{k=0}^{\infty} z_k$ *is* (E, r) *summable to*

$$\sigma \left[x_0 + \sum_{k=1}^{\infty} x_k r^{k-1} \right],$$

where $z_n = \sum_{k=0}^{n} x_k y_{n-k}, n = 0, 1, 2, \ldots.$

The following result is easily proved:

Theorem 6.6 (Limitation theorem) *If* $\sum_{k=0}^{\infty} x_k$ *is* (E, r) *summable, then* $\{x_k\}$ *is bounded.*

Remark 6.2 We now given an example of a bounded, non-convergent sequence which is (E, r) summable. Let $\{s_n\} = \{1, 0, 1, 0, \ldots\}$ and $r = \frac{1}{2}$. Let $\{t_n\}$ be the $(E, \frac{1}{2})$-transform of sequence $\{s_n\}$. Now

$$t_n = \sum_{k=0}^{n} {}^{n}c_k \left(\frac{1}{2}\right)^{n} s_k$$

$$= \frac{1}{2^n} \sum_{k=0}^{n} {}^{n}c_k s_k$$

$$= \frac{1}{2^n} [{}^{n}c_0 + {}^{n}c_2 + {}^{n}c_4 + \cdots + \text{the last term depending on whether } n$$

$$\text{is even or odd}]$$

$$= \frac{1}{2^n} 2^{n-1}$$

$$= \frac{1}{2}, \quad n = 0, 1, 2, \ldots.$$

Thus the sequence $\{1, 0, 1, 0, \ldots\}$ is $(E, \frac{1}{2})$ summable to $\frac{1}{2}$.

Remark 6.3 It was pointed out earlier that the Mazur–Orlicz theorem (Theorem 5.13) fails to hold in the ultrametric case, a counterexample being any regular (N, p_n) method. Theorem 6.6 shows that any (E, r) method is also a counterexample for the failure of the Mazur–Orlicz theorem in the ultrametric set up.

We now have

Theorem 6.7 *Any two Euler methods are consistent.*

Proof Consider the Euler methods (E, r) and (E, s). We then have $|1 - r|, |1 - s| < 1$. Let $\{\sigma_n(r)\}$, $\{\tau_n(s)\}$ be the (E, r), (E, s) transforms of $\{x_k\}$, respectively. Let $\lim_{n \to \infty} \sigma_n(r) = \sigma$ and $\lim_{n \to \infty} \tau_n(s) = \tau$. We claim that $\sigma = \tau$. Now,

$$\sigma_n(r) = (E, r)(\{x_n\})$$

and

$$\tau_n(s) = (E, s)(\{x_n\}).$$

So

$$\sigma_n(r) = (E, r)(E, s)^{-1}(\{\tau_n(s)\})$$

$$= (E, r)\left(E, \frac{1}{s}\right)(\{\tau_n(s)\}), \quad \text{using Corollary 6.1}$$

$$= \left(E, \frac{r}{s}\right)(\{\tau_n(s)\}), \quad \text{using Theorem 6.2.} \tag{6.7}$$

Note that

$$\left|1 - \frac{r}{s}\right| = \left|\frac{s-r}{s}\right| = |s - r|, \quad \text{since } |s| = 1, \text{ using } |1 - s| < 1$$
$$= |(1 - r) - (1 - s)|$$
$$\leq \max(|(1 - r)|, |(1 - s)|)$$
$$< 1,$$

so that $(E, \frac{r}{s})$ is regular, in view of Definition 6.1 and Theorem 6.1. Using (6.7), it follows that $\sigma = \tau$, completing the proof. $\square$

Remark 6.4 In view of Theorem 6.7, we are able to define a parameterless Euler method E of summability as follows:

A sequence $\{x_n\}$ is summable E to σ if there exists $r \in K, |1 - r| < 1$ such that $\{x_n\}$ is (E, r) summable to σ.

Definition 6.2 Let $r \in K$ be such that $|r| < 1$. The Taylor method of order r or the (T, r) method is given by the matrix $(c_{nk}^{(r)})$, $n, k = 0, 1, 2, \ldots$ which is defined as follows:

If $r \neq 0$,

$$c_{nk}^{(r)} = \begin{cases} 0, & k < n; \\ c_n^k r^{k-n}(1 - r)^{n+1}, & k \geq n. \end{cases}$$

If $r = 0$,

$$c_{nk}^{(0)} = \begin{cases} 1, & k = n; \\ 0, & k \neq n. \end{cases}$$

$(c_{nk}^{(r)})$ is called the (T, r) matrix.

Remark 6.5 We note that $r \neq 1$, since $|r| < 1$.

Theorem 6.8 ([1], Theorem 1.6) *Let* $\pi = \sup\{|x|/x \in K, |x| < 1\}$. *Let* $r \in K$ *be such that* $|r| < \pi^{-\frac{1}{\pi - 1}}$. *Then the* (T, r) *method is regular.*

Proof When $r = 0$, the result clearly holds. Now, let $r \neq 0$. Since $c_{nk}^{(r)} = 0, n > k$, it follows that

$$\lim_{n \to \infty} c_{nk}^{(r)} = 0, \quad k = 0, 1, 2, \ldots.$$

Now,

$$\sum_{k=0}^{\infty} c_{nk}^{(r)} = \sum_{k=n}^{\infty} c_{nk}^{(r)}$$

$$= \sum_{k=n}^{\infty} c_k^n r^{k-n} (1-r)^{n+1}$$

$$= (1-r)^{n+1} \sum_{k=n}^{\infty} c_k^n r^{k-n}$$

$$= (1-r)^{n+1}(1-r)^{-(n+1)}, \quad \text{using Binomial theorem}$$

$$\text{and } |n+1| \le 1 \text{ and } |r| < \pi^{-\frac{1}{\pi-1}} \text{ (see [2], p. 51)}$$

$$= 1, \quad n = 0, 1, 2, \dots,$$

so that

$$\lim_{n \to \infty} \sum_{k=0}^{\infty} c_{nk}^{(r)} = 1.$$

When $c_{nk}^{(r)} \neq 0$,

$$|c_{nk}^{(r)}| = |c_k^n r^{k-n}(1-r)^{n+1}|$$

$$\le |r|^{k-n}|1-r|^{n+1}, \quad \text{since } |c_k^n| \le 1$$

$$< |1-r|^{n+1}, \quad \text{since } |r| < 1 \text{ and } k-n \ge 0$$

$$= 1, \quad \text{since } |1-r| = \max(1, |r|) = 1,$$

so that

$$\sup_{n,k} |c_{nk}^{(r)}| < 1. \qquad \bullet$$

Thus the (T, r) method is regular. $\qquad \square$

Remark 6.6 When $K = \mathbb{Q}_p$, it is worthwhile to mention that $\pi = \frac{1}{p}$ and then $\pi^{-\frac{1}{\pi-1}}$ is just the radius of convergence of the exponential.

Notation Let A' denote the transpose of the matrix A.

Theorem 6.9 ([1], Theorem 1.7) *The product of the (T, r) and (T, s) matrices is the matrix* $(1-r)(1-s)(E, (1-r)(1-s))'$.

Proof Let $d_{nk} = (c_{nj}^{(r)})(c_{jk}^{(s)})$. Then $d_{nk} = 0$, if $k < n$ and if $k \geq n$, then

$$
\begin{aligned}
d_{nk} &= \sum_{j=n}^{k} c_{nj}^{(r)} c_{jk}^{(s)} \\
&= \sum_{j=n}^{k} c_j^n r^{j-n} (1-r)^{n+1} c_k^j s^{k-j} (1-s)^{j+1} \\
&= c_k^n \{(1-r)(1-s)\}^{n+1} s^{k-n} \sum_{j=n}^{k} c_{(k-n)}^{(j-n)} \left[\frac{r(1-s)}{s} \right]^{j-n} \\
&= c_k^n \{(1-r)(1-s)\}^{n+1} s^{k-n} \left[1 + \frac{r(1-s)}{s} \right]^{k-n} \\
&= c_k^n \{(1-r)(1-s)\}^{n+1} (s + r - rs)^{k-n}.
\end{aligned}
$$

Let $t = (1-r)(1-s)$ and $k \geq n$. Then

$$
\begin{aligned}
e_{kn}^{(t)} &= c_k^n t^n (1-t)^{k-n} \\
&= c_k^n \{(1-r)(1-s)\}^n \{1 - (1-r)(1-s)\}^{k-n} \\
&= c_k^{(n)} \{(1-r)(1-s)\}^n (s + r - rs)^{k-n}
\end{aligned}
$$

and so

$$
d_{nk} = (1-r)(1-s) e_{kn}^{((1-r)(1-s))},
$$

which completes the proof.   □

Corollary 6.2 *In general, multiplication of matrices is not commutative. However, in the case of (T, r) matrices, it follows from Theorem 6.9 that multiplication is commutative.* •

Corollary 6.3 *The (T, r) matrix is invertible and its inverse is the $\left(T, -\frac{r}{1-r} \right)$ matrix.*

Proof The product of the (T, r), $\left(T, -\frac{r}{1-r} \right)$ matrices is the matrix

$$
\begin{aligned}
&(1-r)\left(1 + \frac{r}{1-r}\right)\left(E, (1-r)\left(1 + \frac{r}{1-r}\right)\right)' \\
&= (1-r)\left(\frac{1}{1-r}\right)\left(E, (1-r)\left(\frac{1}{1-r}\right)\right)' \\
&= (E, 1)'.
\end{aligned}
$$

Since the matrix $(E, 1)'$ is the matrix of the $(E, 1)$ method, which is usual convergence, the result follows.   □

Theorem 6.10 ([1], Theorem 1.10) *If $|s - r| < |1 - r|$, then $(T, r) \subseteq (T, s)$.*

Proof Let $(T, r) = (c_{nk}^{(r)})$, $(T, s) = (c_{nk}^{(s)})$, $\left(T, -\frac{r}{1-r}\right) = (q_{nk})$. Let $\{z_k\}$ be (T, r) summable to y, i.e. $\{\sigma_n\}$ converges to y, where

$$\sigma_n = \sum_{k=n}^{\infty} c_{nk}^{(r)} z_k, \quad n = 0, 1, 2, \ldots.$$

Let $\tau_j = \sum_{k=j}^{\infty} q_{jk}\sigma_k$, $j = 0, 1, 2, \ldots$. Now,

$$\tau_j = \sum_{k=j}^{\infty} q_{jk}\left(\sum_{n=k}^{\infty} c_{kn}^{(r)} z_n\right)$$

$$= \sum_{n=j}^{\infty}\left(\sum_{k=j}^{n} q_{jk} c_{kn}^{(r)}\right) z_n$$

$$= z_j,$$

using Corollary 6.3 of Theorem 6.9 and the fact that convergence is equivalent to unconditional convergence (see [4], p. 133). Now,

$$t_n = \sum_{k=n}^{\infty} c_{nk}^{(s)} z_k$$

$$= \sum_{k=n}^{\infty} c_{nk}^{(s)} \tau_k$$

$$= \sum_{k=n}^{\infty} c_{nk}^{(s)}\left(\sum_{j=k}^{\infty} q_{kj}\sigma_j\right)$$

$$= \sum_{k=n}^{\infty}\sum_{j=k}^{\infty} c_k^n s^{k-n}(1-s)^{n+1}(1-r)^{-(k+1)} c_j^k (-1)^{j-k}\left(\frac{r}{1-r}\right)^{j-k}\sigma_j$$

$$= (1-s)^{n+1}\sum_{j=n}^{\infty} c_j^n (-r)^{j-n}(1-r)^{j+1}\left[\sum_{k=n}^{j} c_{(j-n)}^{(k-n)}\left(-\frac{s}{r}\right)^{k-n}\right]\sigma_j$$

(interchanging order of summation as before)

$$= (1-s)^{n+1}\sum_{j=n}^{\infty} c_j^n (-r)^{j-n}(1-r)^{-(j+1)}\left(1-\frac{s}{r}\right)^{j-n}\sigma_j$$

$$= \left(\frac{1-s}{1-r}\right)^{n+1} \sum_{j=n}^{\infty} c_j^n \left(\frac{s-r}{1-r}\right)^{j-n} \sigma_j,$$

which is the $\left(T, \frac{s-r}{1-r}\right)$ transform of $\{\sigma_j\}$. By hypothesis, $\left|\frac{s-r}{1-r}\right| < 1$ so that the method $\left(T, \frac{s-r}{1-r}\right)$ is regular. Since $\{\sigma_j\}$ converges to y and $\left(T, \frac{s-r}{1-r}\right)$ is regular, $\{t_n\}$ converges to y, i.e. $\{z_k\}$ is (T, s) summable to y, i.e. $(T, r) \subseteq (T, s)$. The proof of the theorem is now complete.   □

References

1. Natarajan, P.N.: Euler and Taylor methods of summability in complete ultrametric fields. J. Anal. **11**, 33–41 (2003)
2. Bachman, G.: Introduction to p-adic Numbers and Valuation Theory. Academic Press, New York (1964)
3. Deepa, R., Natarajan, P.N., Srinivasan, V.: Cauchy multiplication of Euler summable series in ultrametric fields. Comment. Math. Prace Mat. **53**, 73–79 (2013)
4. Van Rooij, A.C.M., Schikhof, W.H.: Non-archimedean analysis. Nieuw Arch. Wisk. **29**, 121–160 (1971)

Chapter 7
Tauberian Theorems

Abstract In this chapter, we prove Tauberian theorems for the Nörlund, the Weighted Mean and the Euler methods.

Keywords The Nörlund method · The Weighted mean method · The Euler method

We recall that when K is a complete, non-trivially valued, ultrametric field, $\lim_{n\to\infty} a_n = 0$ implies that the series $\sum_{n=0}^{\infty} a_n$ converges. Thus $a_n = o(1)$, $n \to \infty$ is a Tauberian condition for any method of summability in ultrametric analysis. Probably, this seems to be the reason for the dearth of meaningful Tauberian theorems in ultrametric analysis, while there is a rich theory of Tauberian theorems in the classical case. However, we now present a few Tauberian theorems in ultrametric analysis, due to Natarajan [1].

Theorem 7.1 ([1], Theorem 1) *Let A be any regular matrix method and $\sum_{n=0}^{\infty} a_n$ be A-summable to s. Let $\lim_{n\to\infty} a_n = \ell$. If $A(n)$ diverges, then $\sum_{n=0}^{\infty} a_n$ converges to s. In other words, $\lim_{n\to\infty} a_n = \ell$ is a Tauberian condition provided $A(n)$ diverges.*

Proof Let $s_n = \sum_{k=0}^{n} a_k$, $n = 0, 1, 2, \ldots$ and $x_n = s_n - n\ell$, $n = 0, 1, 2, \ldots$. Now,

$$
\begin{aligned}
x_n - x_{n-1} &= s_n - n\ell - \{s_{n-1} - (n-1)\ell\} \\
&= (s_n - s_{n-1}) - \ell \\
&= a_n - \ell \\
&\to 0, \quad n \to \infty.
\end{aligned}
$$

So $\{x_n\}$ converges to $\ell^* \in K$ (say), in view of Theorem 1.3.

$$
\text{i.e., } s_n - n\ell - \ell^* \to 0, \quad n \to \infty.
$$

© Springer India 2015

P.N. Natarajan, *An Introduction to Ultrametric Summability Theory*,
Forum for Interdisciplinary Mathematics 2, DOI 10.1007/978-81-322-2559-1_7

Since A is regular,

$$A_n(\{s_k - k\ell - \ell^*\}) \to 0, \quad n \to \infty,$$

i.e., $A_n(\{s_k\}) - \ell A_n(\{k\}) - \ell^* A_n(\{1\}) \to 0, \quad n \to \infty,$

i.e., $s - \ell A_n(\{k\}) - \ell^* \to 0, \quad n \to \infty,$

i.e., $\ell A_n(\{k\}) \to s - \ell^*, \quad n \to \infty.$

Since $A(\{n\})$ diverges by hypothesis, this is possible only if $\ell = 0$. So $\sum_{n=0}^{\infty} a_n$ converges. Since A is regular, $\sum_{n=0}^{\infty} a_n$ converges to s, completing the proof. $\qquad\square$

Corollary 7.1 ([1], p. 299, Corollary) *If $\sum_{n=0}^{\infty} a_n$ is (N, p_n) summable to s, (N, p_n) being regular and if $\lim_{n \to \infty} a_n = \ell$, then $\sum_{n=0}^{\infty} a_n$ converges to s.*

Proof In view of the above Theorem 7.1, it suffices to prove that $\{n\}$ is not summable by the regular (N, p_n) method. Let

$$t_n = \frac{p_0 . n + p_1 . (n-1) + \cdots + p_{n-1} . 1 + p_n . 0}{P_n},$$

$P_n = \sum_{k=0}^{n} p_k, n = 0, 1, 2, \ldots$. Now,

$$|t_{n+1} - t_n| = \left| \frac{p_0 . (n+1) + p_1 . n + \cdots + p_n . 1}{P_{n+1}} - \frac{p_0 . n + p_1 . (n-1) + \cdots + p_{n-1} . 1}{P_n} \right|$$

$$= \left| \frac{P_n\{p_0 + p_1 + \cdots + p_n\} + p_{n+1}\{p_0 . n + p_1 . (n-1) + \cdots + p_{n-1} . 1\}}{P_n P_{n+1}} \right|.$$

Since the valuation of K is ultrametric, $|p_i| < |p_0|$, $i = 1, 2, \ldots$, and $|n| \leq 1$, $n = 1, 2, \ldots$,

$$|P_n| = |P_{n+1}| = |p_0|,$$

$$|P_n\{p_0 + p_1 + \cdots + p_n\}| = |p_0|^2,$$

and

$$|p_{n+1}\{p_0 . n + p_1 . (n-1) + \cdots + p_{n-1} . 1\}| < |p_0|^2$$

so that

$$|t_{n+1} - t_n| = \frac{|p_0|^2}{|p_0|^2} = 1.$$

Consequently, $\{t_n\}$ cannot be Cauchy and so diverges, i.e., $\{n\}$ is not summable (N, p_n). In view of Theorem 7.1, $\sum_{n=0}^{\infty} a_n$ converges to s.   □

Remark 7.1 The hypothesis that $A(\{n\})$ diverges in Theorem 7.1 cannot be dropped. Consider the matrix $A = (a_{nk})$, where

$$a_{nk} = \begin{cases} n, & k = n; \\ 1 - n, & k = n + 1; \\ 0, \text{ otherwise.} \end{cases}$$

We can easily check that A is regular. Now,

$$\sum_{k=0}^{\infty} a_{nk} \cdot k = n \cdot n + (1 - n)(n + 1)$$

$$= n^2 + 1 - n^2$$

$$= 1, \quad n = 0, 1, 2, \ldots,$$

so that $\{n\}$ is transformed (by A) into the sequence $\{1, 1, 1, \ldots\}$. Thus the series $\sum_{n=0}^{\infty} a_n = 1 + 1 + 1 + \cdots$ is A-summable to 1. However, the series $\sum_{n=0}^{\infty} a_n = 1 + 1 + 1 + \cdots$ diverges.

The following result gives an abundance of regular translative methods of summability.

Theorem 7.2 ([1], Theorem 2) *Every regular (N, p_n) method is translative.*

Proof For convenience, we write $A \equiv (N, p_n)$. Now,

$$(A\bar{s})_n \equiv \beta_n = \frac{p_n \cdot 0 + p_{n-1} \cdot s_0 + p_{n-2} \cdot s_1 + \cdots + p_0 \cdot s_{n-1}}{P_n}$$

$$= \frac{P_{n-1}}{P_n} \cdot \frac{p_{n-1} \cdot s_0 + p_{n-2} \cdot s_1 + \cdots + p_0 \cdot s_{n-1}}{P_{n-1}}$$

$$= \frac{P_{n-1}}{P_n} \cdot \alpha_{n-1},$$

where $(As)_n = \alpha_n \cdot \frac{P_{n-1}}{P_n} = \frac{P_n - p_n}{P_n} = 1 - \frac{p_n}{P_n} \to 1, n \to \infty$, since $\left|\frac{p_n}{P_n}\right| = \frac{|p_n|}{|p_0|} \to 0$, $n \to \infty$, the (N, p_n) method being regular. So, if $\alpha_n \to \ell, n \to \infty$, then $\beta_n \to \ell$, $n \to \infty$ too. Again,

$$(As^*)_n \equiv \gamma_n = \frac{p_n \cdot s_1 + p_{n-1} \cdot s_2 + \cdots + p_0 \cdot s_{n+1}}{P_n}$$

$$= \frac{(p_{n+1} \cdot s_0 + p_n \cdot s_1 + \cdots + p_0 \cdot s_{n+1}) - p_{n+1} \cdot s_0}{P_n}$$

$$= \frac{P_{n+1}\alpha_{n+1} - p_{n+1}s_0}{P_n}$$

$$= \frac{P_{n+1}}{P_n} \cdot \alpha_{n+1} - \frac{P_{n+1}}{P_n} \cdot s_0.$$

Since $\frac{P_{n+1}}{P_n} \to 1, n \to \infty$ and $\frac{p_{n+1}}{P_n} \to 0, n \to \infty$, $\alpha_n \to \ell, n \to \infty$ implies that $\gamma_n \to \ell, n \to \infty$, proving that the regular (N, p_n) method is translative. $\qquad\square$

We now prove a Tauberian theorem for regular, translative matrix summability methods.

Theorem 7.3 ([1], Theorem 3) *If* $\sum\limits_{n=0}^{\infty} a_n$ *is summable to s by a regular matrix method*

A *which is translative and* $a_{n+1} - a_n \to \ell, n \to \infty$, *then* $\sum\limits_{n=0}^{\infty} a_n$ *converges to s.*

Proof Let $A = (a_{nk})$ be regular and translative. We first prove that $A(\{n\})$ diverges. If not, i.e., $A(\{n\})$ converges. Since A is translative,

$$\sum_{k=0}^{\infty} a_{nk}(x_{k+1} - x_k) \to 0, \quad n \to \infty,$$

whenever $\{x_k\}$ is A-summable. Since, by assumption, $A(\{n\})$ converges,

$$\sum_{k=0}^{\infty} a_{nk}(k + 1 - k) \to 0, \quad n \to \infty,$$

$$\text{i.e., } \sum_{k=0}^{\infty} a_{nk} \to 0, \quad n \to \infty,$$

which contradicts the fact that

$$\sum_{k=0}^{\infty} a_{nk} \to 1, \quad n \to \infty,$$

this being so since A is regular. Let, now,

$$x_n = a_{n+1} - n\ell, \quad n = 0, 1, 2, \ldots.$$

Then

$$x_{n+1} - x_n = \{a_{n+1} - n\ell\} - \{a_n - (n-1)\ell\}$$
$$= (a_{n+1} - a_n) - \ell$$
$$\to 0, \ n \to \infty.$$

So $\{x_n\}$ converges to ℓ^* (say),

$$\text{i.e., } a_{n+1} - n\ell - \ell^* \to 0, \ n \to \infty,$$

$$\text{i.e., } s_{n+1} - s_n - n\ell - \ell^* \to 0, \ n \to \infty.$$

Since A is regular,

$$A_n(\{s_{k+1} - s_k - k\ell - \ell^*\}) \to 0, \ n \to \infty,$$

$$\text{i.e., } A_n(\{s_{k+1}\}) - A_n(\{s_k\}) - \ell A_n(\{k\}) - \ell^* A_n(\{1\}) \to 0, \ n \to \infty.$$

Since A is translative,

$$s - s - \ell A_n(\{k\}) - \ell^* \to 0, \ n \to \infty.$$

If $\ell \neq 0$, then $A_n(\{k\}) \to -\frac{\ell^*}{\ell}, n \to \infty$, which contradicts the fact that $A(\{n\})$ diverges. So $\ell = 0$. Thus $\{a_n\}$ converges. Now, in view of Theorem 7.1, $\sum\limits_{n=0}^{\infty} a_n$ converges to s, completing the proof of the theorem. $\qquad\qquad\square$

In view of Theorems 7.1 and 7.3, we have the following result.

Theorem 7.4 ([1], Theorem 4) *If $\sum\limits_{n=0}^{\infty} a_n$ is summable by a regular and translative matrix method A, then the Tauberian conditions*

(i) $a_n \to \ell, n \to \infty$;

and

(ii) $a_{n+1} - a_n \to \ell', n \to \infty$

are equivalent.

In the case of regular (N, p_n) methods, we have the following.

Theorem 7.5 ([1], Theorem 5) *If $\sum\limits_{n=0}^{\infty} a_n$ is summable by a regular (N, p_n) method, then the following Tauberian conditions are equivalent:*

(i) $a_n \to \ell, n \to \infty$;

(ii) $\Delta a_n = a_{n+1} - a_n \to \ell', n \to \infty$;

If, further, $a_n \neq 0, n = 0, 1, 2, \ldots$, each of

(iii) $\frac{a_{n+1}}{a_n} \to 1, n \to \infty$;

and

(iv) $\frac{a_{n+2}+a_n}{a_{n+1}} \to 2, n \to \infty$

is a weaker Tauberian condition for the summability of $\sum\limits_{n=0}^{\infty} a_n$ by a regular (N, p_n) method.

Proof The first part follows from Theorem 7.4. We now prove that (iii) implies (iv) and (iv) implies (ii). If (iii) holds, then

$$\frac{a_{n+2} + a_n}{a_{n+1}} = \frac{a_{n+2}}{a_{n+1}} + \frac{a_n}{a_{n+1}}$$
$$\to 2, \quad n \to \infty,$$

so that (iv) holds. Let (iv) hold. Since $\sum\limits_{n=0}^{\infty} a_n$ is (N, p_n) summable, $\{a_n\}$ is bounded, in view of Theorem 5.2 so that there exists $M > 0$ such that $|a_n| \leq M, n = 0, 1, 2, \ldots$. Now,

$$\frac{1}{M}|\Delta^2 a_n| = \frac{1}{M}|a_{n+2} - 2a_{n+1} + a_n|$$
$$\leq \left| \frac{a_{n+2} - 2a_{n+1} + a_n}{a_{n+1}} \right|$$
$$= \left| \frac{a_{n+2} + a_n}{a_{n+1}} - 2 \right|$$
$$\to 0, \quad n \to \infty$$

so that

$$a_{n+2} - 2a_{n+1} + a_n \to 0, \quad n \to \infty,$$

i.e., $(a_{n+2} - a_{n+1}) - (a_{n+1} - a_n) \to 0, \quad n \to \infty.$

Thus $\{a_{n+1} - a_n\}$ is Cauchy and so $a_{n+1} - a_n \to \ell', n \to \infty$ for some $\ell' \in K$ so that (ii) holds. $\square$

Remark 7.2 It is clear that (ii) does not imply (iii) or (iv), as is seen by choosing $a_n = p^n$ in $\mathbb{Q}_p$, for a prime p.

We now prove a Tauberian theorem for weighted mean methods (see [2, 3]).

Theorem 7.6 *If* $\sum\limits_{k=0}^{\infty} a_k$ *is* $(\overline{N}, p_n)$ *summable to* s, *where* $(\overline{N}, p_n)$ *is regular and if*

$$a_n = O\left(\frac{p_n}{P_n}\right), \quad n \to \infty; \tag{7.1}$$

and

$$a_n \to \ell, \quad n \to \infty,$$

then $\sum\limits_{k=0}^{\infty} a_k$ *converges to* s.

Proof We can suppose that $s = 0$. We now claim that $\ell = 0$. If not, choose $\epsilon > 0$ such that $\epsilon < |\ell|$. We can now choose a positive integer N such that

$$|s_n|\left|\frac{p_n}{P_n}\right| < \epsilon, \quad n \geq N, \tag{7.2}$$

using (5.9)

$$|a_n - \ell| < \epsilon, \quad n \geq N,$$

using $\lim_{n\to\infty} a_n = \ell$. Now,

$$|a_n| = |(a_n - \ell) + \ell| = \max(|a_n - \ell|, |\ell|)$$
$$= |\ell|, \quad n \geq N.$$

Also,

$$|a_n| \leq M\left|\frac{p_n}{P_n}\right|, \quad M > 0,$$

in view of (7.1). Thus, for $n \geq N$,

$$|\ell| \leq M\left|\frac{p_n}{P_n}\right|,$$

$$\text{i.e., } \left|\frac{p_n}{P_n}\right| \geq \frac{|\ell|}{M}.$$

Using (7.2), for $n \geq N$,

$$\epsilon > |s_n|\left|\frac{p_n}{P_n}\right| \geq |s_n|\frac{|\ell|}{M},$$

$$\text{i.e., } |s_n| < \frac{\epsilon M}{|\ell|}, \quad n \geq N.$$

In other words, $s_n \to 0$, $n \to \infty$ so that $a_n \to 0$, $n \to \infty$. Thus $\ell = 0$, which is a contradiction. This contradiction leads to the fact that $\ell = 0$. Consequently, $\sum_{k=0}^{\infty} a_k$ converges. Since $(\overline{N}, p_n)$ is regular, $\sum_{k=0}^{\infty} a_k$ converges to 0, completing the proof.   □

We now have the following theorem.

Theorem 7.7 *The (E, r) method is translative.*

Proof Let $\{\sigma_n\}$ be the (E, r) transform of $\{x_k\}$ and $\{\tau_n\}$ be the (E, r) transform of $\{\overline{x}_k\}$. We shall now prove that

$$\sigma_n = \left(1 - \frac{1}{r}\right) \tau_n + \frac{1}{r}\tau_{n+1}, \tag{7.3}$$

i.e., $\displaystyle\sum_{k=0}^{n} {}^{n}c_k r^k (1-r)^{n-k} x_k$

$$= \left(1 - \frac{1}{r}\right) \sum_{k=0}^{n} {}^{n}c_k r^k (1-r)^{n-k}\overline{x}_k$$

$$+ \frac{1}{r} \sum_{k=0}^{n+1} {}^{(n+1)}c_k r^k (1-r)^{n+1-k}\overline{x}_k,$$

i.e., $\displaystyle\sum_{k=0}^{n} {}^{n}c_k r^k (1-r)^{n-k} x_k$

$$= \left(1 - \frac{1}{r}\right) \sum_{k=1}^{n} {}^{n}c_k r^k (1-r)^{n-k} x_{k-1}$$

$$+ \frac{1}{r} \sum_{k=1}^{n+1} {}^{(n+1)}c_k r^k (1-r)^{n+1-k} x_{k-1},$$

$$\text{since } \overline{x}_0 = 0.$$

Now, for $0 \le j \le n$,

$$\text{coefficient of } x_j \text{ on the left side of (7.3)}$$

$$= {}^{n}c_j r^j (1-r)^{n-j};$$

coefficient of x_j on the right side of (7.3)

$$= \left(1 - \frac{1}{r}\right) {}^n c_{(j+1)} r^{j+1} (1 - r)^{n-j-1}$$

$$+ \frac{1}{r} {}^{(n+1)} c_{(j+1)} r^{j+1} (1 - r)^{n-j}$$

$$= -\frac{1-r}{r} {}^n c_{(j+1)} r^{j+1} (1 - r)^{n-j-1}$$

$$+ \frac{1}{r} {}^{(n+1)} c_{(j+1)} r^{j+1} (1 - r)^{n-j}$$

$$= -\frac{1}{r} {}^n c_{(j+1)} r^{j+1} (1 - r)^{n-j}$$

$$+ \frac{1}{r} {}^{(n+1)} c_{(j+1)} r^{j+1} (1 - r)^{n-j}$$

$$= -{}^n c_{(j+1)} r^j (1 - r)^{n-j} + {}^{(n+1)} c_{(j+1)} r^j (1 - r)^{n-j}$$

$$= \left\{ {}^{(n+1)} c_{(j+1)} - {}^n c_{(j+1)} \right\} r^j (1 - r)^{n-j}$$

$$= {}^n c_j r^j (1 - r)^{n-j},$$

thus establishing (7.3). Suppose $\lim_{n\to\infty} \tau_n = s$. Taking limit as $n \to \infty$ in (7.3), we see that

$$\lim_{n\to\infty} \sigma_n = \left(1 - \frac{1}{r}\right) s + \frac{1}{r} s = s,$$

so that (E, r) is right translative. Now,

$$\tau_n = \sum_{k=0}^{n} {}^n c_k r^k (1 - r)^{n-k} \bar{x}_k$$

$$= \sum_{k=1}^{n} {}^n c_k r^k (1 - r)^{n-k} x_{k-1}, \quad \text{since } \bar{x}_0 = 0$$

$$= \sum_{j=0}^{n-1} {}^n c_{(j+1)} r^{j+1} (1 - r)^{n-j-1} x_j$$

$$= \sum_{j=0}^{n-1} {}^n c_{(j+1)} r^{j+1} (1 - r)^{n-j-1} \left\{ \sum_{k=0}^{j} {}^j c_k \left(\frac{1}{r}\right)^k \left(1 - \frac{1}{r}\right)^{j-k} \sigma_k \right\},$$

using Corollary 6.1

$$= \sum_{k=0}^{n-1} r (1 - r)^{n-k-1} \sigma_k \left\{ \sum_{j=k}^{n-1} (-1)^{j-k} {}^n c_{(j+1)} {}^j c_k \right\}.$$

Using the identity

$$\sum_{k=0}^{n-1} \left(\sum_{j=k}^{n-1} (-1)^{j-k} \, {}^n c_{(j+1)}{}^j c_k \right) z^k = \sum_{k=0}^{n-1} z^k,$$

we note that

$$\sum_{j=k}^{n-1} (-1)^{j-k} \, {}^n c_{(j+1)}{}^j c_k = 1, \ \ 0 \le k \le (n-1). \tag{7.4}$$

In view of (7.4), we have

$$\tau_n = \sum_{k=0}^{n-1} r(1-r)^{n-k-1} \sigma_k.$$

Since $|1-r| < 1$, all the conditions of Theorem 4.1 are fulfilled and so

$$\lim_{k \to \infty} \sigma_k = s \text{ implies that } \lim_{n \to \infty} \tau_n = s.$$

Thus (E, r) is left translative. This completes the proof of the theorem. $\qquad \square$

We shall now prove a few Tauberian theorems for the method (E, r) modelled on those proved for (N, p_n) methods by Natarajan [1].

Theorem 7.8 *If $\sum\limits_{k=0}^{\infty} a_k$ is (E, r) summable to σ and if $a_n \to \ell, n \to \infty$, then $\sum\limits_{k=0}^{\infty} a_k$ converges to σ.*

Proof In view of Theorem 7.1, it suffices to prove that the sequence $\{k\}$ of non-negative integers is not (E, r) summable. Let $\{\sigma_n\}$ be the (E, r) transform of $\{k\}$, i.e.,

$$\sigma_n = \sum_{k=0}^{n} {}^n c_k r^k (1-r)^{n-k} k, \ \ n = 0, 1, 2, \ldots.$$

Now,

$$\sigma_{n+1} - \sigma_n = \sum_{k=0}^{n+1} {}^{(n+1)} c_k r^k (1-r)^{n+1-k} k$$

$$- \sum_{k=0}^{n} {}^n c_k r^k (1-r)^{n-k} k$$

$$= \sum_{k=1}^{n+1} {}^{(n+1)} c_k r^k (1-r)^{n+1-k} k$$

$$-\sum_{k=1}^{n} {}^{n}c_k r^k (1-r)^{n-k} k$$

$$= \sum_{k=1}^{n} {}^{(n+1)}c_k r^k (1-r)^{n+1-k} + r^{n+1}(n+1)$$

$$-\sum_{k=1}^{n-1} {}^{n}c_k r^k (1-r)^{n-k} + r^n n.$$

Using $|1-r| < 1, |r| = 1, |k| \le 1, k = 0, 1, 2, \ldots$, we have

$$\left| \sum_{k=1}^{n} {}^{(n+1)}c_k r^k (1-r)^{n+1-k} k \right|$$

$$\le \max_{1 \le k \le n} |{}^{(n+1)}c_k| |r|^k |1-r|^{n+1-k} |k|$$

$$< 1;$$

Similarly,

$$\left| \sum_{k=1}^{n-1} {}^{n}c_k r^k (1-r)^{n-k} k \right| < 1;$$

$$|r^{n+1}(n+1) - r^n n| = |nr^n(r-1) + r^{n+1}|$$

$$= \max\{|n| |r|^n |r-1|, |r|^{n+1}\}$$

$$= 1,$$

so that

$$|\sigma_{n+1} - \sigma_n| = 1, \quad n = 0, 1, 2, \ldots.$$

Thus $\{\sigma_n\}$ is not Cauchy and so diverges, i.e., $\{k\}$ is not (E, r) summable, completing the proof. $\qquad\square$

Using Theorems 7.3 and 7.7, we have the following theorem.

Theorem 7.9 *If* $\sum_{n=0}^{\infty} a_n$ *is* (E, r) *summable to* σ *and if* $a_{n+1} - a_n \to \ell, n \to \infty$,

then $\sum_{n=0}^{\infty} a_n$ *converges to* σ.

As in the case of regular (N, p_n) methods (Theorem 7.5), we have the following result too.

Theorem 7.10 *If* $\sum_{n=0}^{\infty} a_n$ *is* (E, r) *summable, then the following Tauberian conditions are equivalent:*

(i) $a_n \to \ell, n \to \infty;$

(ii) $a_{n+1} - a_n \to \ell', n \to \infty.$

If, further, $a_n \neq 0, n = 0, 1, 2, \ldots,$ *each of*

(iii) $\frac{a_{n+1}}{a_n} \to 1, n \to \infty;$

and

(iv) $\frac{a_{n+2}+a_n}{a_{n+1}} \to 2, n \to \infty$

is a weaker Tauberian condition for the (E, r) *summability of* $\sum\limits_{k=0}^{\infty} a_k.$

References

1. Natarajan, P.N.: Some Tauberian theorems in non-archimedean fields, *p*-adic functional analysis. Lecture Notes in Pure and Applied Mathematics, Marcel Dekker **192**, 297–303 (1997)
2. Natarajan, P.N.: A Tauberian theorem for weighted means in non-archimedean fields, *p*-adic numbers. Ultrametric Anal. Appl. **1**, 368–369 (2009)
3. Natarajan, P.N.: A Tauberian theorem for weighted means in non-archimedean fields—revisited and revised. Comment. Math. Pr. Mat. **54**, 177–178 (2014)

Chapter 8
Silverman-Toeplitz Theorem for Double Sequences and Double Series

Abstract In the present chapter, we introduce double sequences and double series in ultrametric analysis. We prove Silverman-Toeplitz theorem for 4-dimensional infinite matrices. We also prove Schur's and Steinhaus theorems for 4-dimensional matrices.

Keywords Double sequences · Double series · Silverman-Toeplitz theorem · Schur's theorem · Steinhaus theorem

8.1 Double Sequences and Double Series

Natarajan and Srinivasan [1] introduced double sequences and double series in ultrametric analysis and obtained necessary and sufficient conditions for 4-dimensional infinite matrices to be regular. We now briefly introduce these notions.

In the sequel, K denotes a complete, non-trivially valued, ultrametric field.

Definition 8.1 Let $\{x_{m,n}\}$ be a double sequence in K and $x \in K$. We say that $\lim\limits_{m+n \to \infty} x_{m,n} = x$, if for each $\epsilon > 0$, the set $\{(m, n) \in \mathbb{N}^2 : |x_{m,n} - x| \geq \epsilon\}$ is finite, $\mathbb{N}$ being the set of all positive integers. In such a case, x is unique and we say that x is the limit of $\{x_{m,n}\}$.

Definition 8.2 Let $\{x_{m,n}\}$ be a double sequence. We say that

$$s = \sum_{m,n=0}^{\infty} x_{m,n}$$

if

$$s = \lim_{m+n \to \infty} s_{m,n},$$

where

$$s_{m,n} = \sum_{k=0,\ell=0}^{m,n} x_{k,\ell}, \quad m, n = 0, 1, 2, \ldots.$$

© Springer India 2015

P.N. Natarajan, *An Introduction to Ultrametric Summability Theory*,
Forum for Interdisciplinary Mathematics 2, DOI 10.1007/978-81-322-2559-1_8

In such a case, we say that the double series $\sum\limits_{m,n=0}^{\infty} x_{m,n}$ converges to s.

Remark 8.1 If $\lim\limits_{m+n\to\infty} x_{m,n} = x$, then the double sequence $\{x_{m,n}\}$ is bounded.

It is now easy to prove the following results.

Lemma 8.1 $\lim\limits_{m+n\to\infty} x_{m,n} = x$ *if and only if*

(i) $\lim\limits_{n\to\infty} x_{m,n} = x$, $m = 0, 1, 2, \ldots$;

(ii) $\lim\limits_{m\to\infty} x_{m,n} = x$, $n = 0, 1, 2, \ldots$;

and

(iii) *for each $\epsilon > 0$, there exists $N \in \mathbb{N}$ such that $|x_{m,n} - x| < \epsilon$, $m, n \geq N$ which we write as* $\lim\limits_{m,n\to\infty} x_{m,n} = x$.

Lemma 8.2 $\sum\limits_{m,n=0}^{\infty} x_{m,n}$ *converges if and only if*

$$\lim\limits_{m+n\to\infty} x_{m,n} = 0. \qquad (8.1)$$

Definition 8.3 Given, the 4-dimensional infinite matrix $A = (a_{m,n,k,\ell})$ and a double sequence $\{x_{k,\ell}\}$, we define

$$(Ax)_{m,n} = \sum\limits_{k,\ell=0}^{\infty} a_{m,n,k,\ell} x_{k,\ell}, \quad m, n = 0, 1, 2, \ldots,$$

it being assumed that the double series on the right converge. The double sequence $\{(Ax)_{m,n}\}$ is called the A-transform of the double sequence $\{x_{k,\ell}\}$. As in the case of simple sequences, if $\lim\limits_{m+n\to\infty} (Ax)_{m,n} = s$, we say that $\{x_{k,\ell}\}$ is summable A or A-summable to s. If $\lim\limits_{m+n\to\infty} (Ax)_{m,n} = s$ whenever $\lim\limits_{k+\ell\to\infty} x_{k,\ell} = t$, we say that A is convergence preserving. If, further, $s = t$, we say that A is regular.

8.2 Silverman-Toeplitz Theorem

Natarajan and Srinivasan proved the following theorem (see [1]).

Theorem 8.1 (Silverman-Toeplitz theorem) $A = (a_{m,n,k,\ell})$ *is regular if and only if*

$$\lim\limits_{m+n\to\infty} a_{m,n,k,\ell} = 0, \quad k, \ell = 0, 1, 2, \ldots; \qquad (8.2)$$

$$\lim_{m+n\to\infty} \sum_{k,\ell=0}^{\infty} a_{m,n,k,\ell} = 1; \tag{8.3}$$

$$\lim_{m+n\to\infty} \sup_{k\geq0} |a_{m,n,k,\ell}| = 0, \quad \ell = 0, 1, 2, \ldots; \tag{8.4}$$

$$\lim_{m+n\to\infty} \sup_{\ell\geq0} |a_{m,n,k,\ell}| = 0, \quad k = 0, 1, 2, \ldots; \tag{8.5}$$

and

$$\sup_{m,n,k,\ell} |a_{m,n,k,\ell}| < \infty. \tag{8.6}$$

Proof Proof of necessity part

Define the sequence $x = \{x_{k,\ell}\}$ as follows: For any fixed p, q, let

$$x_{k,\ell} = \begin{cases} 1, & \text{when } k = p, \ell = q; \\ 0, & \text{otherwise.} \end{cases}$$

Then

$$(Ax)_{m,n} = a_{m,n,p,q}.$$

Since $\{x_{k,\ell}\}$ has limit 0, it follows that (8.2) is necessary.

Define the sequence $x = \{x_{k,\ell}\}$, where $x_{k,\ell} = 1, k, \ell = 0, 1, 2, \ldots$.
Now,

$$(Ax)_{m,n} = \sum_{k=0,\ell=0}^{\infty,\infty} a_{m,n,k,\ell}, \quad m, n = 0, 1, 2, \ldots.$$

This shows that the series on the right converge and since $\{x_{k,\ell}\}$ has limit 1, it follows that

$$\lim_{m+n\to\infty} \sum_{k=0,\ell=0}^{\infty,\infty} a_{m,n,k,\ell} = 1,$$

so that (8.3) is necessary.

We now show that (8.4) holds. Suppose not. Then, there exists $\ell_0 \in \mathbb{N}$ such that $\lim_{m+n\to\infty} \sup_{k\geq0} |a_{m,n,k,\ell_0}| = 0$ does not hold. So, there exists an $\epsilon > 0$ such that

$$\{(m, n) : \sup_{k\geq0} |a_{m,n,k,\ell_0}| > \epsilon\} \text{ is infinite.} \tag{8.7}$$

Let $m_1 = n_1 = r_1 = 1$. Choose $m_2, n_2 \in \mathbb{N}$ such that $m_2 + n_2 > m_1 + n_1$ and

$$\sup_{0\leq k\leq r_1} |a_{m_2,n_2,k,\ell_0}| < \frac{\epsilon}{8}, \quad \text{using (8.2);}$$

and
$$\sup_{k \geq 0} |a_{m_2,n_2,k,\ell_0}| > \epsilon, \quad \text{using (8.7)}.$$

Then choose $r_2 \in \mathbb{N}$ such that $r_2 > r_1$ and

$$\sup_{k > r_2} |a_{m_2,n_2,k,\ell_0}| < \frac{\epsilon}{8}, \quad \text{using (8.3)}.$$

Inductively choose $m_p + n_p > m_{p-1} + n_{p-1}$ such that

$$\sup_{0 \leq k \leq r_{p-1}} |a_{m_p,n_p,k,\ell_0}| < \frac{\epsilon}{8}; \tag{8.8}$$

$$\sup_{k \geq 0} |a_{m_p,n_p,k,\ell_0}| > \epsilon; \tag{8.9}$$

and then choose $r_p > r_{p-1}$ such that

$$\sup_{k > r_p} |a_{m_p,n_p,k,\ell_0}| < \frac{\epsilon}{8}. \tag{8.10}$$

In view of (8.8), (8.9), (8.10), we have,

$$\sup_{r_{p-1} < k \leq r_p} |a_{m_p,n_p,k,\ell_0}| > \epsilon - \frac{\epsilon}{8} - \frac{\epsilon}{8} = \frac{3\epsilon}{4}.$$

So, there exists k_p, $r_{p-1} < k_p \leq r_p$, such that

$$|a_{m_p,n_p,k_p,\ell_0}| > \frac{3\epsilon}{4}. \tag{8.11}$$

Define the sequence $x = \{x_{k,\ell}\}$ as follows:

$$x_{k,\ell} = \begin{cases} 0, & \text{if } \ell \neq \ell_0; \\ 1, & \text{if } \ell = \ell_0, k = k_p, p = 1, 2, \dots. \end{cases}$$

We note that $\lim_{k+\ell \to \infty} x_{k,\ell} = 0$. Now, in view of (8.8),

$$\left| \sum_{k=0}^{r_{p-1}} a_{m_p,n_p,k,\ell_0} x_{k,\ell_0} \right| \leq \sup_{0 \leq k \leq r_{p-1}} |a_{m_p,n_p,k,\ell_0}| < \frac{\epsilon}{8}; \tag{8.12}$$

Using (8.10), we have,

$$\left| \sum_{k=r_p+1}^{\infty} a_{m_p,n_p,k,\ell_0} x_{k,\ell_0} \right| \leq \sup_{k>r_p} |a_{m_p,n_p,k,\ell_0}| < \frac{\epsilon}{8};$$ (8.13)

and using (8.11), we get,

$$\left| \sum_{k=r_{p-1}+1}^{r_p} a_{m_p,n_p,k,\ell_0} x_{k,\ell_0} \right| = |a_{m_p,n_p,k,\ell_0}| > \frac{3\epsilon}{4}.$$ (8.14)

Thus

$$|(Ax)_{m_p,n_p}| = \left| \sum_{k=0}^{\infty} a_{m_p,n_p,k,\ell_0} x_{k,\ell_0} \right|$$

$$\geq \left| \sum_{k=r_{p-1}+1}^{r_p} a_{m_p,n_p,k,\ell_0} x_{k,\ell_0} \right| - \left| \sum_{k=0}^{r_{p-1}} a_{m_p,n_p,k,\ell_0} x_{k,\ell_0} \right|$$

$$- \left| \sum_{k=r_p+1}^{\infty} a_{m_p,n_p,k,\ell_0} x_{k,\ell_0} \right|$$

$$\geq |a_{m_p,n_p,k_p,\ell_0}| - \sup_{0\leq k\leq r_{p-1}} |a_{m_p,n_p,k,\ell_0}| - \sup_{k>r_p} |a_{m_p,n_p,k,\ell_0}|$$

$$> \frac{3\epsilon}{4} - \frac{\epsilon}{8} - \frac{\epsilon}{8}, \text{ using (8.8), (8.9) and (8.10)}$$

$$= \frac{\epsilon}{2}, \quad p = 1, 2, \ldots.$$

Consequently, $\lim_{m+n\to\infty} (Ax)_{m,n} = 0$ does not hold, which is a contradiction. Thus (8.4) is necessary. The necessity of (8.5) follows in a similar fashion.

To establish (8.6), we shall suppose that (8.6) does not hold and arrive at a contradiction. Since K is non-trivially valued, there exists $\pi \in K$ such that $0 < \rho = |\pi| < 1$. Choose $m_1 = n_1 = 1$. Using (8.2), (8.3), choose $m_2 + n_2 > m_1 + n_1$ such that

$$\sup_{0\leq k+\ell\leq m_1+n_1} |a_{m_2,n_2,k,\ell}| < 2, \text{ using (8.2)};$$

$$\sup_{k+\ell\geq 0} |a_{m_2,n_2,k,\ell}| > \left(\frac{2}{\rho}\right)^6;$$

and

$$\sup_{k+\ell>m_1+n_1} |a_{m_2,n_2,k,\ell}| < 2^2,$$

using (8.3), Lemmas 8.1 and 8.2. It now follows that

$$\sup_{k+\ell>m_2+n_2} |a_{m_2,n_2,k,\ell}| < 2^2.$$

Choose $m_3 + n_3 > m_2 + n_2$ such that

$$\sup_{0\le k+\ell>m_2+n_2} |a_{m_3,n_3,k,\ell}| < 2^2;$$

$$\sup_{k+\ell\ge 0} |a_{m_2,n_2,k,\ell}| > \left(\frac{2}{\rho}\right)^8;$$

and

$$\sup_{k+\ell>m_3+n_3} |a_{m_3,n_3,k,\ell}| < 2^4.$$

Inductively, choose $m_p + n_p > m_{p-1} + n_{p-1}$ such that

$$\sup_{0\le k+\ell\le m_{p-1}+n_{p-1}} |a_{m_p,n_p,k,\ell}| < 2^{p-1}; \tag{8.15}$$

$$\sup_{k+\ell\ge 0} |a_{m_p,n_p,k,\ell}| > \left(\frac{2}{\rho}\right)^{2p+2}; \tag{8.16}$$

and

$$\sup_{k+\ell>m_p+n_p} |a_{m_p,n_p,k,\ell}| < 2^{2p-2}. \tag{8.17}$$

Using (8.15), (8.16), (8.17), we have,

$$\sup_{m_{p-1}+n_{p-1}<k+\ell\le m_p+n_p} |a_{m_p,n_p,k,\ell}|$$

$$> \left(\frac{2}{\rho}\right)^{2p+2} - 2^{2p-2} - 2^{p-1}$$

$$\ge \left(\frac{2}{\rho}\right)^{2p+2} - \left(\frac{2}{\rho}\right)^{2p-2} - \left(\frac{2}{\rho}\right)^{p-1}, \text{ since } \frac{1}{\rho} > 1$$

$$= \left(\frac{2}{\rho}\right)^{p-1} \left[\left(\frac{2}{\rho}\right)^{p+3} - \left(\frac{2}{\rho}\right)^{p-1} - 1\right]$$

$$\ge \left(\frac{2}{\rho}\right)^{p-1} \left[\left(\frac{2}{\rho}\right)^{p+3} - \left(\frac{2}{\rho}\right)^{p-1} - \left(\frac{2}{\rho}\right)^{p-1}\right], \text{ since } \left(\frac{2}{\rho}\right)^{p-1} \ge 1$$

$$= \left(\frac{2}{\rho}\right)^{p-1} \left[\left(\frac{2}{\rho}\right)^4 \left(\frac{2}{\rho}\right)^{p-1} - 2\left(\frac{2}{\rho}\right)^{p-1}\right]$$

$$> \left(\frac{2}{\rho}\right)^{p-1} \left[\left(\frac{2}{\rho}\right)^4 \left(\frac{2}{\rho}\right)^{p-1} - \left(\frac{2}{\rho}\right)\left(\frac{2}{\rho}\right)^{p-1}\right], \text{ since } \frac{2}{\rho} > 2$$

$$= \left(\frac{2}{\rho}\right)^{2p-1} \left[\left(\frac{2}{\rho}\right)^3 - 1\right]$$

$$> \left(\frac{2}{\rho}\right)^{2p-1} [2^3 - 1], \text{ since } \frac{2}{\rho} > 2$$

$$= 7\left(\frac{2}{\rho}\right)^{2p-1}$$

$$> 4\left(\frac{2}{\rho}\right)^{2p-1}$$

$$= \frac{2^{2p+1}}{\rho^{2p-1}}$$

$$> \frac{2^{2p+1}}{\rho^p}, \text{ since } \frac{1}{\rho} > 1. \tag{8.18}$$

Thus, there exists k_p and ℓ_p, $m_{p-1} + n_{p-1} < k_p + \ell_p \le m_p + n_p$ such that

$$|a_{m_p,n_p,k_p,\ell_p}| > \frac{2^{2p+1}}{\rho^p}. \tag{8.19}$$

Now, define the sequence $x = \{x_{k,\ell}\}$ as follows:

$$x_{k,\ell} = \begin{cases} \pi^p, & \text{if } k = k_p, \ell = \ell_p, p = 1, 2, \ldots; \\ 0, & \text{otherwise.} \end{cases}$$

We note that $\lim\limits_{k+\ell \to \infty} x_{k,\ell} = 0$. Now,

$$|(Ax)_{m_p,n_p}| = \left| \sum_{k=0,\ell=0}^{\infty,\infty} a_{m_p,n_p,k,\ell} x_{k,\ell} \right|$$

$$\ge \left| \sum_{k+\ell=(m_{p-1}+n_{p-1})+1}^{m_p+n_p} a_{m_p,n_p,k,\ell} x_{k,\ell} \right| - \left| \sum_{k+\ell=0}^{m_{p-1}+n_{p-1}} a_{m_p,n_p,k,\ell} x_{k,\ell} \right|$$

$$- \left| \sum_{k+\ell=(m_p+n_p)+1}^{\infty} a_{m_p,n_p,k,\ell} x_{k,\ell} \right|$$

$$\ge |a_{m_p,n_p,k_p,\ell_p}||x_{k_p,\ell_p}| - \sup_{0 \le k+\ell \le m_{p-1}+n_{p-1}} |a_{m_p,n_p,k,\ell}|$$

$$- \sup_{m_p+n_p < k+\ell < \infty} |a_{m_p,n_p,k,\ell}|$$

$$> \frac{2^{2p+1}}{\rho^p} \rho^p - 2^{2p-2} - 2^{p-1}, \text{ using (8.15), (8.17) and (8.19)}$$

$$= 2^{2p+1} - 2^{2p-2} - 2^{p-1}$$

$$= 2^{2p-2}(2^3 - 1) - 2^{p-1}$$

$$= 2^{2p-2}(7) - 2^{p-1}$$

$$= 2^{p-1}[7 \cdot 2^{p-1} - 1]$$

$$\geq 2^{p-1}[7 \cdot 2^{p-1} - 2^{p-2}]$$

$$= 2^{p-1}[2^{p-2}(14 - 1)]$$

$$= 2^{p-1}[13 \cdot 2^{p-2}]$$

$$= 13 \cdot 2^{2p-3},$$

i.e., $|(Ax)_{m_p,n_p}| > 13 \cdot 2^{2p-3}, \quad p = 1, 2, \ldots,$

i.e. $\lim\limits_{m+n \to \infty} (Ax)_{m,n} = 0$ does not hold, which is a contradiction. Thus (8.6) is necessary.

Proof of the sufficiency part

Let $x = \{x_{m,n}\}$ be such that $\lim\limits_{m+n \to \infty} x_{m,n} = s$. Then

$$(Ax)_{m,n} - s = \sum_{k=0,\ell=0}^{\infty,\infty} a_{m,n,k,\ell} x_{k,\ell} - s.$$

Using (8.3), we have,

$$\sum_{k=0,\ell=0}^{\infty,\infty} a_{m,n,k,\ell} - r_{m,n} = 1,$$

where

$$\lim_{m+n \to \infty} r_{m,n} = 0. \tag{8.20}$$

Thus

$$(Ax)_{m,n} - s = \sum_{k=0,\ell=0}^{\infty,\infty} a_{m,n,k,\ell}(x_{k,\ell} - s) + r_{m,n}s.$$

Given $\epsilon > 0$, we can choose sufficiently large p, q such that

$$\sup_{k+\ell > p+q} |x_{k,\ell} - s| < \frac{\epsilon}{5H}, \tag{8.21}$$

where $H = \sup\limits_{m,n,k,\ell \geq 0} |a_{m,n,k,\ell}| > 0$. Let $L = \sup\limits_{k,\ell \geq 0} |x_{k,\ell} - s|$.

We now choose $N \in \mathbb{N}$ such that whenever $m + n \geq N$, the following are satisfied.

$$\sup_{0 \leq k+\ell \leq p+q} |a_{m,n,k,\ell}| < \frac{\epsilon}{5pqL}, \text{ using (8.2);} \tag{8.22}$$

$$\sup_{k \geq 0} |a_{m,n,k,\ell}| < \frac{\epsilon}{5qL}, \ell = 0, 1, 2, \ldots, q, \text{ using (8.4);} \tag{8.23}$$

$$\sup_{\ell \geq 0} |a_{m,n,k,\ell}| < \frac{\epsilon}{5pL}, k = 0, 1, 2, \ldots, p, \text{ using (8.5);} \tag{8.24}$$

and

$$|r_{m,n}| < \frac{\epsilon}{5|s|}, \text{ from (8.20).} \tag{8.25}$$

We thus have, whenever $m + n \geq N$,

$$|(Ax)_{m,n} - s| = \left| \sum_{k=0,\ell=0}^{\infty,\infty} a_{m,n,k,\ell}(x_{k,\ell} - s) + r_{m,n}s \right|$$

$$\leq \left| \sum_{k=0,\ell=0}^{p,q} a_{m,n,k,\ell}(x_{k,\ell} - s) \right| + \left| \sum_{k=0,\ell=q+1}^{p,\infty} a_{m,n,k,\ell}(x_{k,\ell} - s) \right|$$

$$+ \left| \sum_{k=p+1,\ell=0}^{\infty,q} a_{m,n,k,\ell}(x_{k,\ell} - s) \right| + \left| \sum_{k=p+1,\ell=q+1}^{\infty,\infty} a_{m,n,k,\ell}(x_{k,\ell} - s) \right|$$

$$+ |r_{m,n}||s|$$

$$< \frac{\epsilon}{5pqL}Lpq + \frac{\epsilon}{5pL}Lp + \frac{\epsilon}{5qL}Lq + \frac{\epsilon}{5H}H + \frac{\epsilon}{5|x|}|x|$$

$$= \epsilon, \text{ using (8.21), (8.22), (8.23), (8.24) and (8.25).}$$

Thus

$$\lim_{m+n \to \infty} (Ax)_{m,n} = s,$$

completing the proof of the theorem. □

8.3 Schur's and Steinhaus Theorems

We now prove Schur's theorem and Steinhaus theorem for 4-dimensional infinite matrices (see [2]).

Definition 8.4 $A = (a_{m,n,k,\ell})$ is called a Schur matrix if $\{(Ax)_{m,n}\} \in c_{ds}$ whenever $x = \{x_{k,\ell}\} \in \ell_{ds}^\infty$, where c_{ds}, ℓ_{ds}^∞, respectively, denote the spaces of convergent double sequences and bounded double sequences.

Definition 8.5 The double sequence $\{x_{m,n}\}$ in K is called a Cauchy sequence if for every $\epsilon > 0$, there exists a positive integer N such that the set

$$\{(m, n), (k, \ell) \in \mathbb{N}^2 : |x_{m,n} - x_{k,\ell}| \geq \epsilon, m, n, k, \ell \geq N\}$$

is finite.

It is now easy to prove the following result.

Theorem 8.2 *The double sequence* $\{x_{m,n}\}$ *in* K *is Cauchy if and only if*

$$\lim_{m+n\to\infty} |x_{m+1,n} - x_{m,n}| = 0;$$

and

$$\lim_{m+n\to\infty} |x_{m,n+1} - x_{m,n}| = 0.$$

Definition 8.6 If every Cauchy double sequence of an ultrametric normed linear space X converges to an element of X, then X is said to be double sequence complete or ds-complete.

For $x = \{x_{m,n}\} \in \ell_{ds}^\infty$, define $\|x\| = \sup_{m,n} |x_{m,n}|$. One can easily prove that ℓ_{ds}^∞ is an ultrametric normed linear space which is ds-complete. With the same definition of norm for elements of c_{ds}, c_{ds} is a closed subspace of ℓ_{ds}^∞. In this context, we recall that Natarajan (Theorems 4.2 and 4.3) proved Schur's theorem and Steinhaus' theorem for 2-dimensional matrices over complete, non-trivially valued, ultrametric fields.

In the sequel, we shall suppose that K is a non-trivially valued ultrametric field which is ds-complete.

Theorem 8.3 (Schur) *The necessary and sufficient conditions for a 4-dimensional infinite matrix* $A = (a_{m,n,k,\ell})$ *to transform double sequences in* ℓ_{ds}^∞ *into double sequences in* c_{ds}, *i.e.* $\{(Ax)_{m,n}\} \in c_{ds}$ *whenever* $x = \{x_{k,\ell}\} \in \ell_{ds}^\infty$ *are:*

$$\lim_{k+\ell\to\infty} a_{m,n,k,\ell} = 0, \quad m, n = 0, 1, 2, \ldots; \tag{8.26}$$

$$\lim_{m+n\to\infty} \sup_{k,\ell\geq 0} |a_{m+1,n,k,\ell} - a_{m,n,k,\ell}| = 0; \tag{8.27}$$

and

$$\lim_{m+n\to\infty} \sup_{k,\ell\geq 0} |a_{m,n+1,k,\ell} - a_{m,n,k,\ell}| = 0. \tag{8.28}$$

Proof Sufficiency. Let (8.26), (8.27), (8.28) hold. Let $x = \{x_{k,\ell}\} \in \ell_{ds}^{\infty}$. We first note that (8.26), (8.27) and (8.28) together imply that (8.6) holds. In view of (8.6) and (8.26),

$$(Ax)_{m,n} = \sum_{k,\ell=0}^{\infty} a_{m,n,k,\ell} x_{k,\ell}, \quad m, n = 0, 1, 2, \ldots$$

is defined, the double series on the right being convergent.

Now,

$$\begin{aligned}
|(Ax)_{m+1,n} - (Ax)_{m,n}| &= \left| \sum_{k,\ell=0}^{\infty} (a_{m+1,n,k,\ell} - a_{m,n,k,\ell}) x_{k,\ell} \right| \\
&\leq M \sup_{k,\ell \geq 0} |a_{m+1,n,k,\ell} - a_{m,n,k,\ell}| \\
&\to 0, \quad m + n \to \infty, \text{ using (8.27),}
\end{aligned}$$

where $|x_{k,\ell}| \leq M, k, \ell = 0, 1, 2, \ldots, M > 0$. Similarly, it follows that

$$|(Ax)_{m,n+1} - (Ax)_{m,n}| \to 0, \quad m + n \to \infty, \text{ using (8.28).}$$

Thus $\{(Ax)_{m,n}\}$ is a Cauchy double sequence in K. Since K is ds-complete, $\{(Ax)_{m,n}\}$ converges and so $\{(Ax)_{m,n}\} \in c_{ds}$, completing the sufficiency part of the proof.

Necessity. Let $\{(Ax)_{m,n}\} \in c_{ds}$ whenever $x = \{x_{k,\ell}\} \in \ell_{ds}^{\infty}$. Consider, the double sequence $\{x_{k,\ell}\}$ where $x_{k,\ell} = 1, k, \ell = 0, 1, 2, \ldots$. $\{x_{k,\ell}\} \in \ell_{ds}^{\infty}$ so that, by hypothesis,

$$(Ax)_{m,n} = \sum_{k,\ell=0}^{\infty} a_{m,n,k,\ell}, \quad m, n = 0, 1, 2, \ldots$$

is defined. Since the series on the right converges, (8.26) holds. Suppose (8.27) does not hold. Then, there exists $\ell_0 \in \mathbb{N}$ such that

$$\lim_{m+n \to \infty} \sup_{k \geq 0} |a_{m+1,n,k,\ell_0} - a_{m,n,k,\ell_0}| = 0$$

does not hold. So, there exists $\epsilon > 0$ such that the set

$$\left\{ (m, n) \in \mathbb{N}^2 : \sup_{k \geq 0} |a_{m+1,n,k,\ell_0} - a_{m,n,k,\ell_0}| > \epsilon \right\}$$

is infinite. Thus, we can choose pairs of integers $m_p, n_p \in \mathbb{N}$ such that $m_1 + n_1 < m_2 + n_2 < \cdots < m_p + n_p < \cdots$ and

$$\sup_{k \geq 0} |a_{m_p+1,n_p,k,\ell_0} - a_{m_p,n_p,k,\ell_0}| > \epsilon, \quad p = 1, 2, \ldots. \tag{8.29}$$

Using (8.26),

$$\lim_{k \to \infty} |a_{m_1+1,n_1,k,\ell_0} - a_{m_1,n_1,k,\ell_0}| = 0.$$

Consequently, there exists $r_1 \in \mathbb{N}$ such that

$$\sup_{k \geq r_1} |a_{m_1+1,n_1,k,\ell_0} - a_{m_1,n_1,k,\ell_0}| < \frac{\epsilon}{2}. \qquad (8.30)$$

Because of (8.29) and (8.30), we have,

$$\sup_{0 \leq k < r_1} |a_{m_1+1,n_1,k,\ell_0} - a_{m_1,n_1,k,\ell_0}| > \epsilon, \qquad (8.31)$$

so that there exists $k_1, 0 \leq k_1 < r_1$ with

$$|a_{m_1+1,n_1,k_1,\ell_0} - a_{m_1,n_1,k_1,\ell_0}| > \epsilon. \qquad (8.32)$$

By hypothesis, (8.2) holds. So we can suppose that

$$\sup_{0 \leq k < r_1} |a_{m_2+1,n_2,k,\ell_0} - a_{m_2,n_2,k,\ell_0}| < \frac{\epsilon}{2}. \qquad (8.33)$$

By (8.29) we have,

$$\sup_{k \geq 0} |a_{m_2+1,n_2,k,\ell_0} - a_{m_2,n_2,k,\ell_0}| > \epsilon. \qquad (8.34)$$

Using (8.26),

$$\lim_{k \to \infty} |a_{m_2+1,n_2,k,\ell_0} - a_{m_2,n_2,k,\ell_0}| = 0$$

so that there exists $r_2 \in \mathbb{N}, r_2 > r_1$ such that

$$\sup_{k \geq r_2} |a_{m_2+1,n_2,k,\ell_0} - a_{m_2,n_2,k,\ell_0}| < \frac{\epsilon}{2}. \qquad (8.35)$$

From (8.33), (8.34), (8.35), we have,

$$\sup_{r_1 \leq k < r_2} |a_{m_2+1,n_2,k,\ell_0} - a_{m_2,n_2,k,\ell_0}| > \epsilon.$$

Thus, there exists $k_2 \in \mathbb{N}$ such that $r_1 \leq k_2 < r_2$ and

$$|a_{m_2+1,n_2,k_2,\ell_0} - a_{m_2,n_2,k_2,\ell_0}| > \epsilon. \qquad (8.36)$$

Inductively, we choose strictly increasing sequences $\{r_p\}, \{k_p\}$ such that $r_{p-1} \leq k_p < r_p$,

$$\sup_{0 \le k < r_{p-1}} |a_{m_p+1,n_p,k,\ell_0} - a_{m_p,n_p,k,\ell_0}| < \frac{\epsilon}{2}; \qquad (8.37)$$

$$\sup_{r_p \le k < \infty} |a_{m_p+1,n_p,k,\ell_0} - a_{m_p,n_p,k,\ell_0}| < \frac{\epsilon}{2}; \qquad (8.38)$$

and

$$|a_{m_p+1,n_p,k_p,\ell_0} - a_{m_p,n_p,k_p,\ell_0}| > \epsilon. \qquad (8.39)$$

Now, define $\{x_{k,\ell}\} \in \ell_{ds}^\infty$, where

$$x_{k,\ell} = \begin{cases} 1, & \text{if } k = k_p, \ell = \ell_0, p = 1, 2, \ldots; \\ 0, & \text{otherwise.} \end{cases}$$

Then,

$$(Ax)_{m_p+1,n_p} - (Ax)_{m_p,n_p} = \sum_{k,\ell=0}^{\infty} (a_{m_p+1,n_p,k,\ell} - a_{m_p,n_p,k,\ell}) x_{k,\ell}$$

$$= \sum_{k=0}^{\infty} (a_{m_p+1,n_p,k,\ell_0} - a_{m_p,n_p,k,\ell_0}) x_{k,\ell_0}$$

$$= \sum_{0 \le k < r_{p-1}} (a_{m_p+1,n_p,k,\ell_0} - a_{m_p,n_p,k,\ell_0}) x_{k,\ell_0}$$

$$\quad + \sum_{r_{p-1} \le k < r_p} (a_{m_p+1,n_p,k,\ell_0} - a_{m_p,n_p,k,\ell_0}) x_{k,\ell_0}$$

$$\quad + \sum_{k \ge r_p} (a_{m_p+1,n_p,k,\ell_0} - a_{m_p,n_p,k,\ell_0}) x_{k,\ell_0}$$

$$= \sum_{0 \le k < r_{p-1}} (a_{m_p+1,n_p,k,\ell_0} - a_{m_p,n_p,k,\ell_0}) x_{k,\ell_0}$$

$$\quad + (a_{m_p+1,n_p,k_p,\ell_0} - a_{m_p,n_p,k_p,\ell_0})$$

$$\quad + \sum_{k \ge r_p} (a_{m_p+1,n_p,k,\ell_0} - a_{m_p,n_p,k,\ell_0}) x_{k,\ell_0}$$

so that

$$(a_{m_p+1,n_p,k_p,\ell_0} - a_{m_p,n_p,k_p,\ell_0}) = \{(Ax)_{m_p+1,n_p} - (Ax)_{m_p,n_p}\}$$

$$\quad - \sum_{0 \le k < r_{p-1}} (a_{m_p+1,n_p,k,\ell_0} - a_{m_p,n_p,k,\ell_0}) x_{k,\ell_0}$$

$$\quad - \sum_{k \ge r_p} (a_{m_p+1,n_p,k,\ell_0} - a_{m_p,n_p,k,\ell_0}) x_{k,\ell_0}.$$

In view of (8.37), (8.38), (8.39), we have,

$$\epsilon < |a_{m_p+1,n_p,k_p,\ell_0} - a_{m_p,n_p,k_p,\ell_0}|$$
$$\leq Max\left[|(Ax)_{m_p+1,n_p} - (Ax)_{m_p,n_p}|, \frac{\epsilon}{2}, \frac{\epsilon}{2}\right],$$

from which it follows that

$$|(Ax)_{m_p+1,n_p} - (Ax)_{m_p,n_p}| > \epsilon, \quad p = 1, 2, \ldots.$$

Consequently $\{(Ax)_{m,n}\} \notin c_{ds}$, a contradiction. Thus (8.27) holds. It can similarly proved that (8.28) holds too. This completes the proof of the theorem.   □

We now deduce the following result.

Theorem 8.4 (Steinhaus) *A 4-dimensional infinite matrix* $A = (a_{m,n,k,\ell})$ *cannot be both a regular and a Schur matrix, i.e. there exists a bounded divergent double sequence which is not A-summable.*

Proof If $A = (a_{m,n,k,\ell})$ is regular, then (8.2) and (8.3) hold. If A were a Schur matrix too, then, $\{a_{m,n,k,\ell}\}_{m,n=0}^{\infty}$ is uniformly Cauchy with respect to $k, \ell = 0, 1, 2, \ldots.$ Since K is ds-complete, $\{a_{m,n,k,\ell}\}_{m,n=0}^{\infty}$ converges uniformly to 0. Consequently we have,

$$\lim_{m+n\to\infty} \sum_{k,\ell=0}^{\infty} a_{m,n,k,\ell} = 0,$$

a contradiction of (8.3), completing the proof of the theorem.

8.4 Some Characterizations of 2-Dimensional Schur Matrices

In this section, we prove some characterizations of 2-dimensional Schur matrices (see [3, 4]).

Theorem 8.5 *The following statements are equivalent:*

(a) $A = (a_{nk}) \in (\ell_{\infty}, c_0);$
(b) (4.16) holds and

$$\lim_{n\to\infty} \sup_{k\geq 0} |a_{nk}| = 0. \tag{8.40}$$

(c) $\lim_{n\to\infty} a_{nk} = 0, k = 0, 1, 2, \ldots$ *and*

$$\lim_{k\to\infty} \sup_{n\geq 0} |a_{nk}| = 0. \tag{8.41}$$

(d)

$$\lim_{n+k\to\infty} a_{nk} = 0. \tag{8.42}$$

Proof In [5, p. 422], Natarajan proved that (a) and (b) are equivalent. We now prove that (b) and (c) are equivalent. Let (b) hold. In view of (8.40), given $\epsilon > 0$, there exists a positive integer N such that

$$\sup_{k\geq 0} |a_{nk}| < \epsilon, \ \ n \geq N. \tag{8.43}$$

Consequently, for fixed $k = 0, 1, 2, \ldots,$

$$|a_{nk}| < \epsilon, \ \ n \geq N$$

so that

$$\lim_{n\to\infty} a_{nk} = 0, \ \ k = 0, 1, 2, \ldots.$$

Now, for $n = 0, 1, 2, \ldots, (N-1)$, using (4.16), we can choose a positive integer U such that

$$\max_{0\leq n\leq N-1} |a_{nk}| < \epsilon, \ \ k \geq U. \tag{8.44}$$

Already,

$$\sup_{n\geq N} |a_{nk}| < \epsilon, \ \ k \geq U, \tag{8.45}$$

using (8.43). Combining (8.44) and (8.45), we have,

$$\sup_{n\geq 0} |a_{nk}| < \epsilon, \ \ k \geq U,$$

so that (8.41) holds. Thus (c) holds, i.e. (b) implies (c). The reverse implication is similarly established.

We shall now prove that (c) and (d) are equivalent. It is clear that (c) implies (d), using Lemma 8.1. Conversely, let (d) hold. Using Lemma 8.1 again,

$$\lim_{n\to\infty} a_{nk} = 0, \ \ k = 0, 1, 2, \ldots;$$

$$\lim_{k\to\infty} a_{nk} = 0, \ \ n = 0, 1, 2, \ldots, \text{i.e., (4.16) holds;}$$

and

$$\lim_{n,k\to\infty} a_{nk} = 0. \tag{8.46}$$

(4.16) along with (8.46) imply (8.41). Thus (c) holds, i.e. (d) implies (c), completing the proof of the theorem. □

Theorem 8.6 *The following statements are equivalent:*

(a) $A = (a_{nk}) \in (\ell_\infty, c)$;
(b) *(i) (4.16) holds;*
  and
  (ii)

$$\lim_{n\to\infty} \sup_{k\geq 0} |a_{n+1,k} - a_{nk}| = 0; \qquad (8.47)$$

(c) *(i) (4.16) holds;*
  (ii) (4.2) holds;
  and
  (iii)

$$\lim_{k\to\infty} \sup_{n\geq 0} |a_{n+1,k} - a_{nk}| = 0; \qquad (8.48)$$

(d) *(i) (4.16) holds;*
  and
  (ii)

$$\lim_{n+k\to\infty} (a_{n+1,k} - a_{nk}) = 0; \qquad (8.49)$$

(e) *(i) (4.16) holds;*
  (ii) (4.2) holds;
  and
  (iii)

$$\lim_{n,k\to\infty} (a_{n+1,k} - a_{nk}) = 0. \qquad (8.50)$$

Proof In [5, pp. 418–421], Natarajan proved the equivalence of (a) and (b). The remaining part of the proof is an easy consequence of Theorem 8.5. For instance, we shall prove the equivalence of (b) and (d). Both in (b) and (d), (4.16) ensures that

$$(Ax)_n = \sum_{k=0}^{\infty} a_{nk}x_k, \quad n = 0, 1, 2, \ldots$$

is defined. Let now (b) hold. Let $B = (b_{nk})$, $b_{nk} = a_{n+1,k} - a_{nk}$, $n, k = 0, 1, 2, \ldots$. Then $B = (b_{nk})$ satisfies (4.16) and (8.40) with a_{nk} replaced by b_{nk}. In view of Theorem 8.5,

$$\lim_{n+k\to\infty} b_{nk} = 0,$$

i.e. $\lim_{n+k\to\infty} (a_{n+1,k} - a_{nk}) = 0.$

Thus (b) implies (d). Converse is similarly proved using Theorem 8.5. The proof of the theorem is now complete. □

References

1. Natarajan, P.N., Srinivasan, V.: Silverman-Toeplitz theorem for double sequences and series and its application to Nörlund means in non-archimedean fields. Ann. Math. Blaise Pascal **9**, 85–100 (2002)
2. Natarajan, P.N.: The Schur and Steinhaus theorems for 4-dimensional matrices in ultrametric fields. Comment. Math. Prace Mat. **51**, 203–209 (2011)
3. Natarajan, P.N.: Some characterizations of Schur matrices in ultrametric fields. Comment. Math. Prace Mat. **52**, 137–142 (2012)
4. Natarajan, P.N.: An addendum to the paper "Some characterizations of Schur matrices in ultrametric fields". Comment. Math. Prace Mat. **53**, 81–82 (2013)
5. Natarajan, P.N.: The Steinhaus theorem for Toeplitz matrices in non-archimedean fields. Comment. Math. Prace Mat. **20**, 417–422 (1978)

Chapter 9
The Nörlund Method and The Weighted Mean Method for Double Sequences

Abstract In the current chapter, we introduce the Nörlund method and the Weighted Mean method for double sequences and establish many of their properties.

Keywords The Nörlund method for double sequences · The Weighted Mean method for double sequences · $\mathcal{N}$-matrix · $\mathcal{N}^{(1)}$ matrix · $\mathcal{N}^{(2)}$ matrix · Convolution product

9.1 The Nörlund Method for Double Sequences

We now introduce the Nörlund mean for double sequences in K.

Definition 9.1 Let $p_{m,n} \in K$ with $|p_{i,j}| < |p_{0,0}|$, $(i, j) \neq (0, 0)$, $i, j = 1, 2, \ldots$. Let

$$P_{m,n} = \sum_{i,j=0}^{m,n} p_{i,j}, \quad m, n = 0, 1, 2, \ldots.$$

Given a double sequence $\{s_{m,n}\}$, we define

$$\sigma_{m,n} = (N, p_{m,n})(\{s_{m,n}\})$$

$$= \frac{\displaystyle\sum_{i,j=0}^{m,n} p_{m-i,n-j} s_{i,j}}{P_{m,n}}, \quad m, n = 0, 1, 2, \ldots.$$

If $\lim_{m+n \to \infty} \sigma_{m,n} = \sigma$, we say that $\{s_{m,n}\}$ is $(N, p_{m,n})$ summable to σ, sometimes written as

$$s_{m,n} \to \sigma(N, p_{m,n}).$$

P.N. Natarajan, *An Introduction to Ultrametric Summability Theory*,
Forum for Interdisciplinary Mathematics 2, DOI 10.1007/978-81-322-2559-1_9

Any double series $\sum\limits_{m,n=0}^{\infty} x_{m,n}$ is said to be $(N, p_{m,n})$ summable to σ if

$$s_{m,n} \to \sigma(N, p_{m,n}),$$

where $s_{m,n} = \sum\limits_{i,j=0}^{m,n} x_{i,j}, m, n = 0, 1, 2, \ldots.$

In view of Theorem 8.1, it is now easy to prove the following result.

Theorem 9.1 *The Nörlund mean* $(N, p_{m,n})$ *is regular if and only if*

$$\lim_{m+n\to\infty} \sup_{0\leq j\leq n} |p_{m-i,n-j}| = 0, \quad 0 \leq i \leq m; \tag{9.1}$$

and

$$\lim_{m+n\to\infty} \sup_{0\leq i\leq m} |p_{m-i,n-j}| = 0, \quad 0 \leq j \leq n; \tag{9.2}$$

In the sequel, let us suppose that $(N, p_{m,n})$, $(N, q_{m,n})$ are regular Nörlund methods such that each row and each column of the 2-dimensional infinite matrices $(p_{m,n})$, $(q_{m,n})$ is a regular Nörlund method for simple sequences. Under this assumption, the following results are proved (for details of proof, see [1]).

Theorem 9.2 *Any two such regular Nörlund methods are consistent.*

Theorem 9.3 $(N, p_{m,n}) \subseteq (N, q_{m,n})$ *if and only if*

$$\lim_{m+n\to\infty} k_{m,n} = 0,$$

where $\{k_{m,n}\}$ *is defined by:*

$$k(x, y) = \sum_{m,n=0}^{\infty} k_{m,n} x^m y^n = \frac{q(x, y)}{p(x, y)},$$

$$p(x, y) = \sum_{m,n=0}^{\infty} p_{m,n} x^m y^n,$$

$$q(x, y) = \sum_{m,n=0}^{\infty} q_{m,n} x^m y^n.$$

Theorem 9.4 $(N, p_{m,n})$ *and* $(N, q_{m,n})$ *are equivalent, i.e.,* $(N, p_{m,n}) \subseteq (N, q_{m,n})$ *and vice versa if and only if*

$$\lim_{m+n\to\infty} k_{m,n} = 0$$

and

$$\lim_{m+n\to\infty} \ell_{m,n} = 0,$$

where $\{\ell_{m,n}\}$ is defined by:

$$\ell(x, y) = \sum_{m,n=0}^{\infty} \ell_{m,n} x^m y^n = \frac{p(x, y)}{q(x, y)}$$

and $\{k_{m,n}\}$ is defined as in Theorem 9.3.

9.2 Cauchy Multiplication of Double Series

Definition 9.2 Let $\{s_{m,n}\}$, $\{t_{m,n}\}$ be two double sequences in K. Then the new sequence $\{z_{m,n}\}$ defined by

$$z_{m,n} = \sum_{k,\ell=0}^{m,n} s_{m-k,n-\ell} t_{k,\ell}, \quad m, n = 0, 1, 2, \ldots$$

is called the Cauchy product of $\{s_{m,n}\}$ and $\{t_{m,n}\}$.

We now prove the following result which is the analogue of Theorem 4.16 for double series.

Theorem 9.5 Let $\sum_{m,n=0}^{\infty} s_{m,n}$, $\sum_{m,n=0}^{\infty} t_{m,n}$ be two infinite double series with terms in K. Then $\sum_{m,n=0}^{\infty} z_{m,n}$, where

$$z_{m,n} = \sum_{k,\ell=0}^{m,n} s_{m-k,n-\ell} t_{k,\ell}, \quad m, n = 0, 1, 2, \ldots,$$

converges for every convergent series $\sum_{m,n=0}^{\infty} t_{m,n}$ if and only if $\sum_{m,n=0}^{\infty} s_{m,n}$ converges.

Further, if $\sum_{m,n=0}^{\infty} s_{m,n}$, $\sum_{m,n=0}^{\infty} t_{m,n}$ converge, then $\sum_{m,n=0}^{\infty} z_{m,n}$ converges and

$$\sum_{m,n=0}^{\infty} z_{m,n} = \left(\sum_{m,n=0}^{\infty} s_{m,n} \right) \left(\sum_{m,n=0}^{\infty} t_{m,n} \right).$$

Proof Let $\sum\limits_{m,n=0}^{\infty} s_{m,n}$ be given such that $\sum\limits_{m,n=0}^{\infty} z_{m,n}$ converges for every convergent

double series $\sum\limits_{m,n=0}^{\infty} t_{m,n}$. By hypothesis, $\sum\limits_{m,n=0}^{\infty} z_{m,n}$ converges for the convergent

double series $\sum\limits_{m,n=0}^{\infty} t_{m,n}$, where $t_{0,0} = 1$, $t_{m,n} = 0$, otherwise. In this case, $z_{m,n} =$

$s_{m,n}$ so that $\sum\limits_{m,n=0}^{\infty} s_{m,n}$ converges.

Conversely, let $\sum\limits_{m,n=0}^{\infty} s_{m,n}$, $\sum\limits_{m,n=0}^{\infty} t_{m,n}$ converge. So $\lim\limits_{m+n\to\infty} s_{m,n} = 0$, $\lim\limits_{m+n\to\infty}$
$t_{m,n} = 0$. Thus, there exists $M > 0$ such that $|s_{m,n}| < M$, $|t_{m,n}| < M$ for all $m, n =$
$0, 1, 2, \ldots$. Given $\epsilon > 0$, choose positive integers M_1, N_1 such that $|s_{m,n}| < \frac{\epsilon}{M}$,
$|t_{m,n}| < \frac{\epsilon}{M}$ for all $m > M_1, n > N_1$. Since $\lim\limits_{m+n\to\infty} s_{m-k,n-\ell} = 0$ for every fixed
$k, \ell = 0, 1, 2, \ldots$, we can choose positive integers $M_2 > M_1, N_2 > N_1$ such that
for all $m > M_2, n > N_2$,

$$\sup_{\substack{0\le k\le M_1 \\ 0\le \ell\le N_1}} |s_{m-k,n-\ell}| < \frac{\epsilon}{M};$$

$$\sup_{\substack{0\le k\le M_1 \\ N_1+1\le \ell\le n}} |s_{m-k,n-\ell}| < \frac{\epsilon}{M};$$

and

$$\sup_{\substack{M_1+1\le k\le m \\ 0\le \ell\le N_1}} |s_{m-k,n-\ell}| < \frac{\epsilon}{M}.$$

Then, for every $m > M_2, n > N_2$,

$$|z_{m,n}| = \left| \sum_{k,\ell=0}^{m,n} s_{m-k,n-\ell} t_{k,\ell} \right|$$

$$= \left| \sum_{\substack{0\le k\le M_1 \\ 0\le \ell\le N_1}} s_{m-k,n-\ell} t_{k,\ell} + \sum_{\substack{0\le k\le M_1 \\ \ell>N_1}} s_{m-k,n-\ell} t_{k,\ell} \right.$$

$$\left. + \sum_{\substack{k>M_1 \\ 0\le \ell\le N_1}} s_{m-k,n-\ell} t_{k,\ell} + \sum_{\substack{k>M_1 \\ \ell>N_1}} s_{m-k,n-\ell} t_{k,\ell} \right|$$

$$\leq \max \left[\sup_{\substack{0 \leq k \leq M_1 \\ 0 \leq \ell \leq N_1}} |s_{m-k,n-\ell} t_{k,\ell}|, \right.$$

$$\sup_{\substack{0 \leq k \leq M_1 \\ N_1+1 \leq \ell \leq n}} |s_{m-k,n-\ell} t_{k,\ell}|,$$

$$\sup_{\substack{M_1+1 \leq k \leq m \\ 0 \leq \ell \leq N_1}} |s_{m-k,n-\ell} t_{k,\ell}|,$$

$$\left. \sup_{\substack{M_1+1 \leq k \leq m \\ N_1+1 \leq \ell \leq n}} |s_{m-k,n-\ell} t_{k,\ell}| \right]$$

$$\leq \max \left[\frac{\epsilon}{M} \cdot M, \frac{\epsilon}{M} \cdot M, \frac{\epsilon}{M} \cdot M, \frac{\epsilon}{M} \cdot M \right]$$

$$= \epsilon.$$

Thus $\sum_{m,n=0}^{\infty} z_{m,n}$ converges. Now, we will prove that

$$\sum_{m,n=0}^{\infty} z_{m,n} = \left(\sum_{m,n=0}^{\infty} s_{m,n} \right) \left(\sum_{m,n=0}^{\infty} t_{m,n} \right).$$

Now,

$$\left(\sum_{m,n=0}^{\infty} s_{m,n} \right) \left(\sum_{m,n=0}^{\infty} t_{m,n} \right) = \lim_{m+n \to \infty} \left(\sum_{k,\ell=0}^{m,n} s_{k,\ell} \right) \lim_{m+n \to \infty} \left(\sum_{k,\ell=0}^{m,n} t_{k,\ell} \right)$$

$$= \lim_{m+n \to \infty} \left(\sum_{k,\ell=0}^{m,n} s_{k,\ell} \right) \left(\sum_{k,\ell=0}^{m,n} t_{k,\ell} \right)$$

$$= \lim_{m+n \to \infty} \left(\sum_{k,\ell=0}^{m,n} \left(\sum_{i,j=0}^{k,\ell} s_{k-i,\ell-j} t_{i,j} \right) \right)$$

$$= \lim_{m+n \to \infty} \sum_{k,\ell=0}^{m,n} z_{k,\ell}$$

$$= \sum_{m,n=0}^{\infty} z_{m,n},$$

completing the proof of the theorem.   □

We need the following definitions in the sequel (see [2]).

Definition 9.3 Any matrix $A = (a_{m,n,k,\ell})$ for which $\sup\limits_{m,n,k,\ell} |a_{m,n,k,\ell}| < \infty$ is called an $\mathcal{N}$ matrix.

Definition 9.4 Any matrix $A = (a_{m,n,k,\ell})$ is called an $\mathcal{N}^{(1)}$ matrix if for $m = 0, 1, 2, \ldots$, there exists a positive integer k_m such that

$$a_{m,n,k,\ell} = 0, \quad k > k_m, n, \ell = 0, 1, 2, \ldots.$$

Definition 9.5 An infinite matrix $A = (a_{m,n,k,\ell})$ is called an $\mathcal{N}^{(2)}$ matrix if for $n = 0, 1, 2, \ldots$, there exists a positive integer ℓ_n such that

$$a_{m,n,k,\ell} = 0, \quad \ell > \ell_n, m, k = 0, 1, 2, \ldots.$$

Definition 9.6 Given two infinite matrices $A = (a_{m,n,k,\ell})$, $B = (b_{m,n,k,\ell})$, m, n, $k, \ell = 0, 1, 2, \ldots$, their convolution product is defined as the matrix $C = (c_{m,n,k,\ell})$, where, for $k, \ell = 0, 1, 2, \ldots$

$$c_{m,n,k,\ell} = \sum_{i,j=0}^{k,\ell} a_{m,n,k-i,\ell-j} b_{m,n,i,j},$$

$m, n = 0, 1, 2, \ldots$. In such a case, we write $C = A * B$.

Theorem 9.6 *If $A = (a_{m,n,k,\ell})$, $B = (b_{m,n,k,\ell})$ are both $\mathcal{N}^{(1)}$ or both $\mathcal{N}^{(2)}$ or both $\mathcal{N}$, then, their convolution C is also $\mathcal{N}^{(1)}$ or $\mathcal{N}^{(2)}$ or $\mathcal{N}$, respectively. Further, for $m, n = 0, 1, 2, \ldots$*

$$\sum_{k,\ell=0}^{\infty} c_{m,n,k,\ell} = \left(\sum_{k,\ell=0}^{\infty} a_{m,n,k,\ell} \right) \left(\sum_{k,\ell=0}^{\infty} b_{m,n,k,\ell} \right),$$

provided the series on the right converge.

Proof Let $A = (a_{m,n,k,\ell})$, $B = (b_{m,n,k,\ell})$ be $\mathcal{N}^{(1)}$. Since A is $\mathcal{N}^{(1)}$, for $m = 0, 1, 2, \ldots$, there exists a positive integer $k_m^{(1)}$ such that $a_{m,n,k,\ell} = 0$ for $k > k_m^{(1)}$, $n, \ell = 0, 1, 2, \ldots$. Also, since B is $\mathcal{N}^{(1)}$, there exists a positive integer $k_m^{(2)}$ such that $b_{m,n,k,\ell} = 0$ for $k > k_m^{(2)}$, $n, \ell = 0, 1, 2, \ldots$. Let $k_m = \max(k_m^{(1)}, k_m^{(2)})$. Then,

$$a_{m,n,k,\ell} = 0 = b_{m,n,k,\ell} \text{ for } k > k_m, n, \ell = 0, 1, 2, \ldots.$$

Let now, $k > 2k_m$. Then for $n, \ell = 0, 1, 2, \ldots$,

$$c_{m,n,k,\ell} = \sum_{i,j=0}^{k,\ell} a_{m,n,k-i,\ell-j} b_{m,n,i,j} = 0.$$

Thus C is an $\mathcal{N}^{(1)}$ matrix. Similarly, we can prove that if A, B are both $\mathcal{N}^{(2)}$, then C is also $\mathcal{N}^{(2)}$.

Now, let A, B be $\mathcal{N}$ matrices.

Then

$$\sup_{m,n,k,\ell} |a_{m,n,k,\ell}| < \infty;$$

and

$$\sup_{m,n,k,\ell} |b_{m,n,k,\ell}| < \infty.$$

Thus

$$\sup_{m,n,k,\ell} |c_{m,n,k,\ell}| = \sup_{m,n,k,\ell} \left| \sum_{i,j=0}^{k,\ell} a_{m,n,k-i,\ell-j} b_{m,n,i,j} \right|$$

$$\leq \sup_{m,n,k,\ell} \left[\max_{\substack{0 \leq i \leq k \\ 0 \leq j \leq \ell}} |a_{m,n,k-i,\ell-j} b_{m,n,i,j}| \right]$$

$$< \infty.$$

Consequently, C is an $\mathcal{N}$ matrix too. The remaining part of the theorem follows from the latter part of Theorem 9.5. □

If $A = (a_{m,n,k,\ell})$, $B = (b_{m,n,k,\ell})$ are both regular, it is worthwhile to check whether the convolution of A and B is also regular or otherwise. We leave the details to the reader.

9.3 The Weighted Mean Method for Double Sequences

We now introduce weighted means for double sequences and extend theorems dealing with weighted means for simple sequences (for details, refer to [3]).

Definition 9.7 The $(\overline{N}, p_{m,n})$ method is defined by the infinite matrix $(a_{m,n,k,\ell})$, where

$$a_{m,n,k,\ell} = \begin{cases} \dfrac{p_{k,\ell}}{P_{m,n}}, & \text{if } k \leq m, \ell \leq n; \\ 0, & \text{if } k > m \text{ or } \ell > n, \end{cases}$$

$$P_{m,n} = \sum_{i,j=0}^{m,n} p_{i,j}, \quad m, n = 0, 1, 2, \ldots,$$

with the sequence $\{p_{m,n}\}$ of weights satisfying the conditions

$$p_{m,n} \neq 0, \quad m, n = 0, 1, 2, \ldots, \tag{9.3}$$

for each fixed pair (i, j),

$$\begin{aligned}|p_{k,\ell}| \leq |P_{i,j}|, \; k &= 0, 1, 2, \ldots, i; \\ i &= 0, 1, 2, \ldots; \\ \ell &= 0, 1, 2, \ldots, j; \\ j &= 0, 1, 2, \ldots. \end{aligned} \tag{9.4}$$

Remark 9.1 From (9.4), it is clear that for every fixed $i = 0, 1, 2, \ldots$,

$$\begin{aligned}|p_{i,\ell}| \leq |P_{i,j}|, \; \ell &= 0, 1, 2, \ldots, j; \\ j &= 0, 1, 2, \ldots \end{aligned} \tag{9.5}$$

and for every fixed $j = 0, 1, 2, \ldots$,

$$\begin{aligned}|p_{k,j}| \leq |P_{i,j}|, \; k &= 0, 1, 2, \ldots, i; \\ i &= 0, 1, 2, \ldots. \end{aligned} \tag{9.6}$$

Note that (9.4) is equivalent to

$$\max_{\substack{0 \leq k \leq i \\ 0 \leq \ell \leq j}} |p_{k,\ell}| \leq |P_{i,j}|, \; i, j = 0, 1, 2, \ldots; \tag{9.7}$$

(9.5) is equivalent to

$$\max_{0 \leq \ell \leq j} |p_{i,\ell}| \leq |P_{i,j}|, \; j = 0, 1, 2, \ldots; \tag{9.8}$$

(9.6) is equivalent to

$$\max_{0 \leq k \leq i} |p_{k,j}| \leq |P_{i,j}|, \; i = 0, 1, 2, \ldots. \tag{9.9}$$

Since the valuation is ultrametric,

$$|P_{i,j}| \leq \max_{\substack{0 \leq k \leq i \\ 0 \leq \ell \leq j}} |p_{k,\ell}|, \; i, j = 0, 1, 2, \ldots. \tag{9.10}$$

Combining (9.7) and (9.10), we have for every fixed pair (i, j),

$$|P_{i,j}| = \max_{\substack{0 \leq k \leq i \\ 0 \leq \ell \leq j}} |p_{k,\ell}|, \; i, j = 0, 1, 2, \ldots. \tag{9.11}$$

Using (9.11), we have,

$$P_{m,n} \neq 0, \quad m, n = 0, 1, 2, \ldots. \tag{9.12}$$

Remark 9.2 (9.11) implies

$$|P_{m+1,n+1}| \geq |P_{m,n}|; \tag{9.13}$$

$$|P_{m,n+1}| \geq |P_{m,n}|; \tag{9.14}$$

and

$$|P_{m+1,n}| \geq |P_{m,n}|. \tag{9.15}$$

Proof From (9.11),

$$
\begin{aligned}
|P_{m+1,n+1}| &= \max_{\substack{0 \leq k \leq m+1 \\ 0 \leq \ell \leq n+1}} |p_{k,\ell}| \\
&= \max \left[\max_{\substack{0 \leq k \leq m \\ 0 \leq \ell \leq n}} |p_{k,\ell}|, |p_{m,n+1}|, |p_{m+1,n}|, |p_{m+1,n+1}| \right] \\
&= \max \left[|P_{m,n}|, |p_{m,n+1}|, |p_{m+1,n}|, |p_{m+1,n+1}| \right] \\
&\geq |P_{m,n}|.
\end{aligned}
$$

In a similar fashion, we can prove that (9.11) implies (9.14) and (9.15). □

Theorem 9.7 $(\overline{N}, p_{m,n})$ *is regular if and only if*

$$\lim_{m+n \to \infty} |P_{m,n}| = \infty; \tag{9.16}$$

$$\lim_{m+n \to \infty} \frac{\max\limits_{0 \leq k \leq m} |p_{k,\ell}|}{|P_{m,n}|} = 0, \quad \ell = 0, 1, 2, \ldots; \tag{9.17}$$

and

$$\lim_{m+n \to \infty} \frac{\max\limits_{0 \leq \ell \leq n} |p_{k,\ell}|}{|P_{m,n}|} = 0, \quad k = 0, 1, 2, \ldots. \tag{9.18}$$

Proof Necessity part. Let $(\overline{N}, p_{m,n})$ be regular. Using (8.2),

$$\lim_{m+n \to \infty} |a_{m,n,0,0}| = 0,$$

i.e., $\quad \lim\limits_{m+n \to \infty} \left| \dfrac{P_{0,0}}{P_{m,n}} \right| = 0.$

Thus,

$$\lim_{m+n\to\infty} |P_{m,n}| = \infty,$$

since $p_{0,0} \neq 0$, using (9.3).

Also, using (8.4), for $\ell = 0, 1, 2, \ldots$,

$$\lim_{m+n\to\infty} \sup_{k\geq 0} |a_{m,n,k,\ell}| = 0,$$

i.e., $\quad \lim_{m+n\to\infty} \max_{0\leq k\leq m} \frac{|p_{k,\ell}|}{|P_{m,n}|} = 0, \quad \ell = 0, 1, 2, \ldots,$

i.e., $\quad \lim_{m+n\to\infty} \frac{\max\limits_{0\leq k\leq m} |p_{k,\ell}|}{|P_{m,n}|} = 0, \quad \ell = 0, 1, 2, \ldots.$

Thus (9.17) holds. In the same manner, we can prove (9.18).

Sufficiency part. Let (9.16), (9.17), (9.18) hold. For every fixed $k, \ell = 0, 1, 2, \ldots$,

$$\lim_{m+n\to\infty} a_{m,n,k,\ell} = \lim_{m+n\to\infty} \frac{p_{k,\ell}}{P_{m,n}} = 0,$$

in view of (9.16). Now,

$$\lim_{m+n\to\infty} \sum_{k,\ell=0}^{\infty} a_{m,n,k,\ell} = \lim_{m+n\to\infty} \frac{\sum\limits_{k,\ell=0}^{m,n} p_{k,\ell}}{P_{m,n}}$$

$$= \lim_{m+n\to\infty} \frac{P_{m,n}}{P_{m,n}} = 1, \quad \text{using (9.12)}.$$

Also, for every fixed $\ell = 0, 1, 2, \ldots$,

$$\lim_{m+n\to\infty} \sup_{k\geq 0} |a_{m,n,k,\ell}| = \lim_{m+n\to\infty} \sup_{0\leq k\leq m} |a_{m,n,k,\ell}|$$

$$= \lim_{m+n\to\infty} \frac{\max\limits_{0\leq k\leq m} |p_{k,\ell}|}{|P_{m,n}|}$$

$$= 0, \quad \text{using (9.17)}.$$

In a similar fashion, we can prove that for $k = 0, 1, 2, \ldots, \quad \lim_{m+n\to\infty} \sup_{\ell\geq 0} |a_{m,n,k,\ell}| = 0$,

using (9.18). Finally, $|a_{m,n,k,\ell}| = 0$, if $k > m$ or $\ell > n$; However, if $k \leq m, \ell \leq n$,

$$|a_{m,n,k,\ell}| = \frac{|p_{k,\ell}|}{|P_{m,n}|}$$

$$\leq 1, \quad \text{in view of (9.4)}.$$

Thus, $\sup\limits_{m,n,k,\ell} |a_{m,n,k,\ell}| < \infty$. Consequently, the method $(\overline{N}, p_{m,n})$ is regular. $\qquad\square$

Example 9.1 Let $K = \mathbb{Q}_p$, the p-adic field for a prime p. Then, $0 < c = |p|_p < 1$, $|\cdot|_p$ denoting the p-adic valuation.

Let

$$p_{m,n} = \begin{cases} p^{m+n}, & \text{if } m+n \text{ is odd;} \\ \frac{1}{p^{m+n}}, & \text{if } m+n \text{ is even,} \end{cases} \qquad (9.19)$$

and

$$s_{m,n} = \begin{cases} \frac{1}{p^{m+n}}, & \text{if } m+n \text{ is odd;} \\ p^{m+n}, & \text{if } m+n \text{ is even.} \end{cases} \qquad (9.20)$$

It is clear that $\{s_{m,n}\}$ does not converge. Let $\{t_{m,n}\}$ be the $(\overline{N}, p_{m,n})$ transform of $\{s_{m,n}\}$, i.e.

$$t_{m,n} = \sum_{k,\ell=0}^{\infty} a_{m,n,k,\ell} s_{k,\ell}, \quad m,n = 0,1,2,\ldots,$$

where

$$a_{m,n,k,\ell} = \begin{cases} \frac{p_{k,\ell}}{P_{m,n}}, & \text{if } k \le m, \ell \le n; \\ 0, & \text{if } k > m \text{ or } \ell > n, \end{cases}$$

$$P_{m,n} = \sum_{i,j=0}^{m,n} p_{i,j}, \quad m,n = 0,1,2,\ldots.$$

Now,

$$|t_{2k,2\ell+1}|_p = \left| \sum_{i,j=0}^{2k,2\ell+1} \frac{p_{i,j} s_{i,j}}{P_{2k,2\ell+1}} \right|_p$$

$$= \frac{|2k(2\ell+1)|_p}{|P_{2k,2\ell+1}|_p}.$$

If k, ℓ are both odd or both even, then

$$|t_{2k,2\ell+1}|_p = \frac{|2k(2\ell+1)|_p}{|P_{2k,2\ell+1}|_p}$$

$$= \frac{|2k(2\ell+1)|_p}{\left| 1 + 2p + \frac{3}{p^2} + \cdots + (k+\ell+1)\frac{1}{p^{k+1}} \right.}$$
$$\left. + (k+\ell+1)p^{k+\ell+1} + \cdots + 2 \cdot \frac{1}{p^{2k+2\ell}} + 1 \cdot p^{2k+2\ell+1} \right|_p}$$

$$= \left| \frac{2k(2\ell + 1)}{2 \cdot \frac{1}{p^{2k+2\ell}}} \right|_p$$

$$\leq |p^{2k+2\ell}|_p$$

$$= |p|_p^{2(k+\ell)};$$

If k is even and ℓ is odd or k is odd and ℓ is even, then

$$|t_{2k,2\ell+1}|_p = \frac{|2k(2\ell + 1)|_p}{|P_{2k,2\ell+1}|_p}$$

$$= \frac{|2k(2\ell + 1)|_p}{\left| 1 + 2p + \frac{3}{p^2} + \cdots + (k + \ell + 1)p^{k+\ell} \right. } \\ \left. + (k + \ell + 1)\frac{1}{p^{k+\ell+1}} + \cdots + 2 \cdot \frac{1}{p^{2k+2\ell}} + 1 \cdot p^{2k+2\ell+1} \right|_p$$

$$= \left| \frac{2k(2\ell + 1)}{\frac{2}{p^{2k+2\ell}}} \right|_p$$

$$\leq |p|_p^{2(k+\ell)};$$

Similarly, we can prove that

$$|t_{2k+1,2\ell}|_p \leq |p|_p^{2(k+\ell)};$$

$$|t_{2k,2\ell}|_p \leq |p|_p^{2(k+\ell)}$$

and

$$|t_{2k+1,2\ell+1}|_p \leq |p|_p^{2(k+\ell+1)}.$$

It now follows that $\lim_{m+n\to\infty} t_{m,n} = 0$. Thus, $\{s_{m,n}\}$, through non-convergent, is summable $(\overline{N}, p_{m,n})$ to 0.

Now, we will prove that the $(\overline{N}, p_{m,n})$ method corresponding to $\{p_{m,n}\}$ defined in (9.19) is regular.

Case 1 When $m + n$ is even;

By definition and using (9.11),

$$\lim_{m+n\to\infty} |P_{m,n}|_p = \lim_{m+n\to\infty} \max_{\substack{0 \leq i \leq m \\ 0 \leq j \leq n}} |p_{i,j}|_p$$

$$= \lim_{m+n\to\infty} \left| \frac{1}{p^{m+n}} \right|_p$$

$$= \lim_{m+n\to\infty} \frac{1}{|p|_p^{m+n}}$$

$$= \infty.$$

Also, for $\ell = 0, 1, 2, \ldots,$

$$\lim_{m+n\to\infty} \frac{\max_{0\leq i\leq m} |p_{i,\ell}|_p}{|P_{m,n}|_p} = \lim_{m+n\to\infty} \frac{\max_{0\leq i\leq m} |p_{i,\ell}|_p}{\max_{\substack{0\leq i\leq m \\ 0\leq j\leq n}} |p_{i,j}|_p}$$

$$= \lim_{m+n\to\infty} \frac{\max_{0\leq i\leq m} |p_{i,\ell}|_p}{\frac{1}{|p|_p^{m+n}}}$$

$$= \lim_{m+n\to\infty} |p^{m+n}|_p \max_{0\leq i\leq m} |p_{i,\ell}|_p.$$

Subcase (1) If both m, ℓ are even or both are odd, then

$$\lim_{m+n\to\infty} \frac{\max_{0\leq i\leq m} |p_{i,\ell}|_p}{|P_{m,n}|_p} = \lim_{m+n\to\infty} |p^{m+n}|_p \left|\frac{1}{p^{m+\ell}}\right|_p$$

$$= \lim_{m+n\to\infty} |p^{n-\ell}|_p$$

$$= \lim_{m+n\to\infty} |p|_p^{n-\ell}$$

$$= 0;$$

Subcase (2) When m is even and ℓ is odd or vice versa, then

$$\lim_{m+n\to\infty} \frac{\max_{0\leq i\leq m} |p_{i,\ell}|_p}{|P_{m,n}|_p} = \lim_{m+n\to\infty} |p|_p^{m+n} \max_{0\leq i\leq m} |p_{i,\ell}|_p$$

$$= \lim_{m+n\to\infty} |p^{m+n}|_p \left|\frac{1}{p^{m+\ell-1}}\right|_p$$

$$= \lim_{m+n\to\infty} |p^{n-\ell+1}|_p$$

$$= \lim_{m+n\to\infty} |p|_p^{n-\ell+1}$$

$$= 0.$$

Case 2 When $m+n$ is odd. This case can be discussed in a similar fashion. (9.18) can be proved similarly. Consequently, $(\overline{N}, p_{m,n})$, where $\{p_{m,n}\}$ is defined as in (9.19), is regular.

Theorem 9.8 (Limitation theorem) *If* $\{s_{m,n}\}$ *is* $(\overline{N}, p_{m,n})$ *summable to s, then*

$$s_{m,n} - s = o\left(\frac{P_{m,n}}{p_{m,n}}\right), \quad m+n \to \infty, \tag{9.21}$$

in the sense that

$$\frac{p_{m,n}}{P_{m,n}}(s_{m,n} - s) \to 0, \quad m+n \to \infty.$$

Proof If $\{t_{m,n}\}$ is the $(\overline{N}, p_{m,n})$ transform of $\{s_{m,n}\}$, then

$$\left|\frac{p_{m,n}}{P_{m,n}}(s_{m,n} - s)\right| = \left|\frac{\begin{array}{c}(P_{m,n}t_{m,n} - P_{m,n-1}t_{m,n-1} - P_{m-1,n}t_{m-1,n} + P_{m-1,n-1}t_{m-1,n-1}) \\ -(P_{m,n} - P_{m,n-1} - P_{m-1,n} + P_{m-1,n-1})s\end{array}}{P_{m,n}}\right|$$

$$= \left|\begin{array}{c}(t_{m,n} - s) - \frac{P_{m,n-1}}{P_{m,n}}(t_{m,n-1} - s) - \frac{P_{m-1,n}}{P_{m,n}}(t_{m-1,n} - s) \\ + \frac{P_{m-1,n-1}}{P_{m,n}}(t_{m-1,n-1} - s)\end{array}\right|$$

$$\leq \max\left[\begin{array}{c}|t_{m,n} - s|, \left|\frac{P_{m,n-1}}{P_{m,n}}\right||t_{m,n-1} - s|, \left|\frac{P_{m-1,n}}{P_{m,n}}\right||t_{m-1,n} - s|, \\ \left|\frac{P_{m-1,n-1}}{P_{m,n}}\right||t_{m-1,n-1} - s|\end{array}\right]$$

$$\leq \max[|t_{m,n} - s|, |t_{m,n-1} - s|, |t_{m-1,n} - s|, |t_{m-1,n-1} - s|]$$

in view of Remark 9.2. Since $\displaystyle\lim_{m+n\to\infty} t_{m,n} = s$, it follows that

$$\lim_{m+n\to\infty}\left|\frac{p_{m,n}}{P_{m,n}}(s_{m,n} - s)\right| = 0,$$

i.e. $s_{m,n} - s = o\left(\dfrac{P_{m,n}}{p_{m,n}}\right), \quad m+n \to \infty,$

completing the proof of the theorem.　　　　　　　　　　　　　□

Definition 9.8 We say that $(\overline{N}, p_{m,n})$ is included in $(\overline{N}, q_{m,n})$ (or $(\overline{N}, q_{m,n})$ includes $(\overline{N}, p_{m,n})$), written as

$$(\overline{N}, p_{m,n}) \subseteq (\overline{N}, q_{m,n})$$

if $s_{m,n} \to s(\overline{N}, p_{m,n})$ implies $s_{m,n} \to s(\overline{N}, q_{m,n})$.

We now prove a pair of inclusion theorems involving weighted means for double sequences.

Theorem 9.9 (Comparison theorem for two weighted means for double sequences)
Let $(\overline{N}, p_{m,n})$, $(\overline{N}, q_{m,n})$ be two weighted mean methods such that

$$q_{m,n} = O(p_{m,n}), \quad m+n \to \infty \tag{9.22}$$

in the sense that there exists $M > 0$ such that

$$\left|\frac{q_{m,n}}{p_{m,n}}\right| \leq M, \quad m, n = 0, 1, 2, \ldots$$

and

$$P_{m,n} = o(Q_{m,n}), \quad m+n \to \infty \tag{9.23}$$

in the sense that

$$\left| \frac{P_{m,n}}{Q_{m,n}} \right| \to 0, \quad m+n \to \infty,$$

where $P_{m,n} = \sum\limits_{k,\ell=0}^{m,n} p_{k,\ell}, \ Q_{m,n} = \sum\limits_{k,\ell=0}^{m,n} q_{k,\ell}, \ m, n = 0, 1, 2, \ldots.$

Then

$$(\overline{N}, p_{m,n}) \subseteq (\overline{N}, q_{m,n}).$$

Proof Using (9.22), (9.23), $\lim\limits_{m+n\to\infty} \left| \dfrac{q_{m,n} P_{m,n}}{p_{m,n} Q_{m,n}} \right| = 0$ and so there exists $H > 0$ such that

$$\left| \frac{q_{m,n} P_{m,n}}{p_{m,n} Q_{m,n}} \right| \le H, \quad m, n = 0, 1, 2, \ldots. \tag{9.24}$$

Using (9.22), there exists $M > 0$ such that

$$\left| \frac{q_{m,n}}{p_{m,n}} \right| \le M, \quad m, n = 0, 1, 2, \ldots. \tag{9.25}$$

Let, for a double sequence $\{s_{m,n}\}$,

$$t_{m,n} = \frac{\sum\limits_{i,j=0}^{m,n} p_{i,j} s_{i,j}}{P_{m,n}},$$

$$u_{m,n} = \frac{\sum\limits_{i,j=0}^{m,n} q_{i,j} s_{i,j}}{Q_{m,n}}, \quad m, n = 0, 1, 2, \ldots.$$

Then,

$$p_{0,0} s_{0,0} = P_{0,0} t_{0,0};$$
$$p_{m,n} s_{m,n} = P_{m,n} t_{m,n} - P_{m-1,n} t_{m-1,n} - P_{m,n-1} t_{m,n-1} + P_{m-1,n-1} t_{m-1,n-1},$$

so that

$$s_{m,n} = \frac{P_{m,n} t_{m,n} - P_{m-1,n} t_{m-1,n} - P_{m,n-1} t_{m,n-1} + P_{m-1,n-1} t_{m-1,n-1}}{p_{m,n}},$$

$$s_{m,0} = \frac{P_{m,0} t_{m,0} - P_{m-1,0} t_{m-1,0}}{p_{m,0}},$$

$$s_{0,n} = \frac{P_{0,n} t_{0,n} - P_{0,n-1} t_{0,n-1}}{p_{0,n}},$$

where we suppose that

$$P_{-1,n} = 0, \quad P_{m,-1} = 0, \quad P_{-1,-1} = 0.$$

Now,

$$u_{m,n} = \frac{1}{Q_{m,n}} \left(\sum_{i,j=0}^{m,n} q_{i,j} s_{i,j} \right)$$

$$= \frac{1}{Q_{m,n}} \sum_{i,j=0}^{m,n} q_{i,j} \left\{ \frac{P_{i,j} t_{i,j} - P_{i-1,j} t_{i-1,j} - P_{i,j-1} t_{i,j-1} + P_{i-1,j-1} t_{i-1,j-1}}{p_{i,j}} \right\}$$

$$= \sum_{k,\ell=0}^{\infty} c_{m,n,k,\ell} t_{k,\ell},$$

where,

$$c_{m,n,k,\ell} = \begin{cases} \left(\dfrac{q_{k,\ell}}{p_{k,\ell}} - \dfrac{q_{k,\ell+1}}{p_{k,\ell+1}} - \dfrac{q_{k+1,\ell}}{p_{k+1,\ell}} + \dfrac{q_{k+1,\ell+1}}{p_{k+1,\ell+1}} \right) \dfrac{P_{k,\ell}}{Q_{m,n}}, & \text{if } k < m, \ell < n; \\[4mm] \left(\dfrac{q_{k,\ell}}{p_{k,\ell}} - \dfrac{q_{k,\ell+1}}{p_{k,\ell+1}} \right) \dfrac{P_{k,\ell}}{Q_{m,n}}, & \text{if } k = m, \ell < n; \\[4mm] \left(\dfrac{q_{k,\ell}}{p_{k,\ell}} - \dfrac{q_{k+1,\ell}}{p_{k+1,\ell}} \right) \dfrac{P_{k,\ell}}{Q_{m,n}}, & \text{if } k < m, \ell = n; \\[4mm] \dfrac{q_{k,\ell}}{p_{k,\ell}} \dfrac{P_{k,\ell}}{Q_{k,\ell}}, & \text{if } k = m, \ell = n; \\[4mm] 0, & \text{if } k > m \text{ or } \ell > n. \end{cases}$$

Using (9.22), (9.23), we get

$$\lim_{m+n \to \infty} c_{m,n,k,\ell} = 0, \quad k, \ell = 0, 1, 2, \dots.$$

If $s_{m,n} = 1, m, n = 0, 1, 2, \dots$, then

$$t_{m,n} = \frac{\displaystyle\sum_{i,j=0}^{m,n} p_{i,j} s_{i,j}}{P_{m,n}}$$

$$= \frac{\displaystyle\sum_{i,j=0}^{m,n} p_{i,j}}{P_{m,n}}$$

$$= \frac{P_{m,n}}{P_{m,n}}$$

$$= 1, \quad \text{using (9.12).}$$

Similarly, $u_{m,n} = 1$, $m, n = 0, 1, 2, \ldots$ so that

$$u_{m,n} = \sum_{k,\ell=0}^{\infty} c_{m,n,k,\ell} t_{k,\ell}$$

which yields

$$1 = \sum_{k,\ell=0}^{\infty} c_{m,n,k,\ell}(1),$$

i.e., $\sum_{k,\ell=0}^{\infty} c_{m,n,k,\ell} = 1$, $m, n = 0, 1, 2, \ldots$.

Consequently,

$$\lim_{m+n \to \infty} \left(\sum_{k,\ell=0}^{\infty} c_{m,n,k,\ell} \right) = 1.$$

If $k < m$ and $\ell < n$, then

$$
|c_{m,n,k,\ell}| = \left| \left(\frac{q_{k,\ell}}{p_{k,\ell}} - \frac{q_{k,\ell+1}}{p_{k,\ell+1}} - \frac{q_{k+1,\ell}}{p_{k+1,\ell}} + \frac{q_{k+1,\ell+1}}{p_{k+1,\ell+1}} \right) \frac{P_{k,\ell}}{Q_{m,n}} \right|
$$

$$
\leq \max \left[\left| \frac{q_{k,\ell}}{p_{k,\ell}} \right| \left| \frac{P_{k,\ell}}{Q_{m,n}} \right|, \left| \frac{q_{k,\ell+1}}{p_{k,\ell+1}} \right| \left| \frac{P_{k,\ell}}{Q_{m,n}} \right|, \left| \frac{q_{k+1,\ell}}{p_{k+1,\ell}} \right| \left| \frac{P_{k,\ell}}{Q_{m,n}} \right|, \left| \frac{q_{k+1,\ell+1}}{p_{k+1,\ell+1}} \right| \left| \frac{P_{k,\ell}}{Q_{m,n}} \right| \right]
$$

$$
\leq \max \left[\left| \frac{q_{k,\ell}}{p_{k,\ell}} \right| \left| \frac{P_{k,\ell}}{Q_{k,\ell}} \right|, \left| \frac{q_{k,\ell+1}}{p_{k,\ell+1}} \right| \left| \frac{P_{k,\ell+1}}{Q_{k,\ell+1}} \right|, \left| \frac{q_{k+1,\ell}}{p_{k+1,\ell}} \right| \left| \frac{P_{k+1,\ell}}{Q_{k+1,\ell}} \right|, \left| \frac{q_{k+1,\ell+1}}{p_{k+1,\ell+1}} \right| \left| \frac{P_{k+1,\ell+1}}{Q_{k+1,\ell+1}} \right| \right]
$$

$$
\leq H, \quad \text{by (9.24)},
$$

since $k < m$, $\ell < n$ imply

$$|Q_{k,\ell}|, |Q_{k,\ell+1}|, |Q_{k+1,\ell}|, |Q_{k+1,\ell+1}| \leq |Q_{m,n}|$$

and so

$$\frac{1}{|Q_{m,n}|} \leq \frac{1}{|Q_{k,\ell}|}, \frac{1}{|Q_{k,\ell+1}|}, \frac{1}{|Q_{k+1,\ell}|}, \frac{1}{|Q_{k+1,\ell+1}|}$$

and

$$|P_{k,\ell}| \leq |P_{k+1,\ell}|, |P_{k,\ell+1}|, |P_{k+1,\ell+1}|.$$

The cases when $k = m$, $\ell < n$ and $k < m$, $\ell = n$ can be similarly discussed. If $k = m$, $\ell = n$, then

$$
|c_{m,n,m,n}| = \left| \begin{matrix} q_{m,n} & P_{m,n} \\ p_{m,n} & Q_{m,n} \end{matrix} \right|
$$

$$
\leq H, \quad \text{by (9.24)}
$$

and

$$|c_{m,n,k,\ell}| = 0 \leq H \text{ if } k > m \text{ or } \ell > n.$$

Thus

$$\sup_{m,n,k,\ell} |c_{m,n,k,\ell}| < \infty.$$

We shall now prove that for every fixed $\ell = 0, 1, 2, \ldots,$

$$\lim_{m+n\to\infty} \sup_{k\geq 0} |c_{m,n,k,\ell}| = 0.$$

For every fixed $\ell = 0, 1, 2, \ldots,$ there are three cases, viz., $\ell < n,$ $\ell = n$ and $\ell > n.$

Case 1 When $\ell < n,$

$$\lim_{m+n\to\infty} \sup_{k\geq 0} |c_{m,n,k,\ell}| = \lim_{m+n\to\infty} \max_{0\leq k\leq m} |c_{m,n,k,\ell}|$$

$$= \lim_{m+n\to\infty} \left[\max \left(\max_{0\leq k\leq m-1} |c_{m,n,k,\ell}|, |c_{m,n,m,\ell}| \right) \right]$$

$$= \lim_{m+n\to\infty} \max \left[\max_{0\leq k\leq m-1} \left| \left(\frac{q_{k,\ell}}{p_{k,\ell}} - \frac{q_{k,\ell+1}}{p_{k,\ell+1}} - \frac{q_{k+1,\ell}}{p_{k+1,\ell}} + \frac{q_{k+1,\ell+1}}{p_{k+1,\ell+1}} \right) \frac{P_{k,\ell}}{Q_{m,n}} \right|, \right.$$

$$\left. \left| \left(\frac{q_{m,\ell}}{p_{m,\ell}} - \frac{q_{m,\ell+1}}{p_{m,\ell+1}} \right) \frac{P_{m,\ell}}{Q_{m,n}} \right| \right]$$

$$\leq \lim_{m+n\to\infty} M \left| \frac{P_{m,n}}{Q_{m,n}} \right|, \text{ using (9.25)}$$

$$= 0, \text{ using (9.23)}.$$

Therefore,

$$\lim_{m+n\to\infty} \sup_{k\geq 0} |c_{m,n,k,\ell}| = 0 \text{ if } \ell < n.$$

Case 2 The case $\ell = n$ can be proved similarly.
Case 3 If $\ell > n,$ then

$$c_{m,n,k,\ell} = 0, \quad m, n = 0, 1, 2, \ldots,$$

by definition. Thus $\lim_{m+n\to\infty} \sup_{k\geq 0} |c_{m,n,k,\ell}| = 0.$

Consequently,

$$\lim_{m+n\to\infty} \sup_{k\geq 0} |c_{m,n,k,\ell}| = 0.$$

Similarly, we can prove that

$$\lim_{m+n\to\infty} \sup_{\ell\geq 0} |c_{m,n,k,\ell}| = 0, \quad k = 0, 1, 2, \ldots.$$

The method $(c_{m,n,k,\ell})$ is thus regular and so

$$(\overline{N}, p_{m,n}) \subseteq (\overline{N}, q_{m,n}),$$

completing the proof of the theorem. □

Theorem 9.10 (Comparison theorem for a $(\overline{N}, p_{m,n})$ method and a regular matrix)
Let $(\overline{N}, p_{m,n})$ be a weighted mean method and $A = (a_{m,n,k,\ell})$ be a regular matrix. If

$$\lim_{k+\ell \to \infty} \frac{a_{m,n,k,\ell}}{p_{k,\ell}} P_{k,\ell} = 0, \quad m, n = 0, 1, 2, \ldots \tag{9.26}$$

and

$$P_{m,n} = O(p_{m,n}), \quad m + n \to \infty,$$

i.e., $\left| \dfrac{P_{m,n}}{p_{m,n}} \right| \leq M, \quad M > 0, m, n = 0, 1, 2, \ldots, \tag{9.27}$

then

$$(\overline{N}, p_{m,n}) \subseteq A.$$

Proof Let $\{s_{m,n}\}$ be any double sequence, $\{t_{m,n}\}$, $\{\tau_{m,n}\}$ be its $(\overline{N}, p_{m,n})$, A-transforms, respectively. Then,

$$t_{m,n} = \frac{\displaystyle\sum_{i,j=0}^{m,n} p_{i,j} s_{i,j}}{P_{m,n}},$$

$$\tau_{m,n} = \sum_{k,\ell=0}^{\infty} a_{m,n,k,\ell} s_{k,\ell}, \quad m, n = 0, 1, 2, \ldots.$$

Now,

$$s_{m,n} = \frac{P_{m,n} t_{m,n} - P_{m-1,n} t_{m-1,n} - P_{m,n-1} t_{m,n-1} + P_{m-1,n-1} t_{m-1,n-1}}{p_{m,n}},$$

$m, n = 0, 1, 2, \ldots,$ where

$$t_{-1,n} = t_{m,-1} = t_{-1,-1} = 0.$$

Let $\lim\limits_{m+n\to\infty} t_{m,n} = s$. Then,

$$
\begin{aligned}
\tau_{m,n} &= \sum_{k,\ell=0}^{\infty} a_{m,n,k,\ell} s_{k,\ell} \\
&= \sum_{k,\ell=0}^{\infty} \frac{a_{m,n,k,\ell}}{p_{k,\ell}} \left[P_{k,\ell} t_{k,\ell} - P_{k-1,\ell} t_{k-1,\ell} - P_{k,\ell-1} t_{k,\ell-1} + P_{k-1,\ell-1} t_{k-1,\ell-1} \right] \\
&= \sum_{k,\ell=0}^{\infty} \left[\frac{a_{m,n,k,\ell}}{p_{k,\ell}} - \frac{a_{m,n,k+1,\ell}}{p_{k+1,\ell}} - \frac{a_{m,n,k,\ell+1}}{p_{k,\ell+1}} + \frac{a_{m,n,k+1,\ell+1}}{p_{k+1,\ell+1}} \right] P_{k,\ell} t_{k,\ell},
\end{aligned}
$$

$$(9.28)$$

noting that $\tau_{m,n}$ exists, $m, n = 0, 1, 2, \ldots$, since the right-hand side of (9.28) exists using (9.26) and using the fact that $\{t_{k,\ell}\}$ is convergent and so bounded and

$$
\left| \frac{P_{k,\ell}}{P_{k,\ell+1}} \right| \le 1, \quad \left| \frac{P_{k,\ell}}{P_{k+1,\ell}} \right| \le 1 \text{ and } \left| \frac{P_{k,\ell}}{P_{k+1,\ell+1}} \right| \le 1.
$$

We now write

$$
\tau_{m,n} = \sum_{k,\ell=0}^{\infty} b_{m,n,k,\ell} t_{k,\ell},
$$

where

$$
b_{m,n,k,\ell} = \left(\frac{a_{m,n,k,\ell}}{p_{k,\ell}} - \frac{a_{m,n,k+1,\ell}}{p_{k+1,\ell}} - \frac{a_{m,n,k,\ell+1}}{p_{k,\ell+1}} + \frac{a_{m,n,k+1,\ell+1}}{p_{k+1,\ell+1}} \right) P_{k,\ell}.
$$

By using (9.27), $\sup\limits_{m,n,k,\ell} |a_{m,n,k,\ell}| < \infty$ and the fact that

$$
\left| \frac{P_{k,\ell}}{P_{k,\ell+1}} \right| \le 1, \quad \left| \frac{P_{k,\ell}}{P_{k+1,\ell}} \right| \le 1 \text{ and } \left| \frac{P_{k,\ell}}{P_{k+1,\ell+1}} \right| \le 1,
$$

we have,

$$
\sup_{m,n,k,\ell} |b_{m,n,k,\ell}| < \infty.
$$

Also, using (9.27) and the regularity of A, we get

$$
\lim_{m+n\to\infty} b_{m,n,k,\ell} = 0, \quad k, \ell = 0, 1, 2, \ldots.
$$

Let $s_{m,n} = 1, m, n = 0, 1, 2, \ldots$. Then $t_{m,n} = 1, m, n = 0, 1, 2, \ldots$. It now follows that

$$\sum_{k,\ell=0}^{\infty} b_{m,n,k,\ell} = \sum_{k,\ell=0}^{\infty} a_{m,n,k,\ell}, \quad m, n = 0, 1, 2, \ldots.$$

Consequently,

$$\lim_{m+n\to\infty} \left(\sum_{k,\ell=0}^{\infty} b_{m,n,k,\ell} \right) = \lim_{m+n\to\infty} \left(\sum_{k,\ell=0}^{\infty} a_{m,n,k,\ell} \right)$$

$$= 1,$$

since A is regular. Now, for $\ell = 0, 1, 2, \ldots,$

$$\lim_{m+n\to\infty} \sup_{k\geq 0} |b_{m,n,k,\ell}| = \lim_{m+n\to\infty} \sup_{k\geq 0} \left| \left(\frac{a_{m,n,k,\ell}}{p_{k,\ell}} - \frac{a_{m,n,k+1,\ell}}{p_{k+1,\ell}} - \frac{a_{m,n,k,\ell+1}}{p_{k,\ell+1}} \right. \right.$$

$$\left. \left. + \frac{a_{m,n,k+1,\ell+1}}{p_{k+1,\ell+1}} \right) P_{k,\ell} \right|$$

$$\leq \lim_{m+n\to\infty} \sup_{k\geq 0} \left[\max \left(\left| \frac{a_{m,n,k,\ell}}{p_{k,\ell}} P_{k,\ell} \right|, \left| \frac{a_{m,n,k+1,\ell}}{p_{k+1,\ell}} P_{k,\ell} \right|, \right. \right.$$

$$\left. \left. \left| \frac{a_{m,n,k,\ell+1}}{p_{k,\ell+1}} P_{k,\ell} \right|, \left| \frac{a_{m,n,k+1,\ell+1}}{p_{k+1,\ell+1}} P_{k,\ell} \right| \right) \right].$$

Now,

$$\lim_{m+n\to\infty} \sup_{k\geq 0} \left| \frac{a_{m,n,k,\ell}}{p_{k,\ell}} P_{k,\ell} \right| = \lim_{m+n\to\infty} \sup_{k\geq 0} |a_{m,n,k,\ell}| \left| \frac{P_{k,\ell}}{p_{k,\ell}} \right|$$

$$\leq M \lim_{m+n\to\infty} \sup_{k\geq 0} |a_{m,n,k,\ell}|, \quad \text{using (9.27)}$$

$$= 0,$$

since A is regular and so

$$\lim_{m+n\to\infty} \sup_{k\geq 0} |a_{m,n,k,\ell}| = 0.$$

Again,

$$\lim_{m+n\to\infty} \sup_{k\geq 0} \left| \frac{a_{m,n,k+1,\ell}}{p_{k+1,\ell}} P_{k,\ell} \right| \leq \lim_{m+n\to\infty} \sup_{k\geq 0} |a_{m,n,k+1,\ell}| \left| \frac{P_{k+1,\ell}}{p_{k+1,\ell}} \right|$$

$$\leq M \lim_{m+n\to\infty} \sup_{k\geq 0} |a_{m,n,k+1,\ell}|, \quad \text{using (9.27)}$$

$$= 0, \quad \text{since } A \text{ is regular.}$$

Similarly, we can prove that

$$\lim_{m+n\to\infty} \sup_{k\geq 0} \left| \frac{a_{m,n,k,\ell+1}}{p_{k,\ell+1}} P_{k,\ell} \right| = 0$$

and

$$\lim_{m+n\to\infty} \sup_{k\geq 0} \left| \frac{a_{m,n,k+1,\ell+1}}{p_{k+1,\ell+1}} P_{k,\ell} \right| = 0.$$

Consequently,

$$\lim_{m+n\to\infty} \sup_{k\geq 0} |b_{m,n,k,\ell}| = 0, \quad \ell = 0, 1, 2, \ldots.$$

Similarly we can prove that

$$\lim_{m+n\to\infty} \sup_{\ell\geq 0} |b_{m,n,k,\ell}| = 0, \quad k = 0, 1, 2, \ldots.$$

The method $(b_{m,n,k,\ell})$ is thus regular and so $\lim_{m+n\to\infty} t_{m,n} = s$ implies that $\lim_{m+n\to\infty} \tau_{m,n} = s$. In other words,

$$(\overline{N}, p_{m,n}) \subseteq A,$$

completing the proof of the theorem.   □

We conclude, this book by remarking that there are other aspects of ultrametric analysis concerning summability theory, viz., sequence spaces, matrix transformations between sequence spaces, etc. We have not included these aspects in the current edition. For such topics, interested readers can refer to [4–20].

References

1. Natarajan, P.N., Srinivasan, V.: Silverman-Toeplitz theorem for double sequences and series and its application to Nörlund means in non-archimedean fields. Ann. Math. Blaise Pascal **9**, 85–100 (2002)
2. Natarajan, P.N., Sakthivel, S.:Multiplication of double series and convolution of double infinite matrices in non-archimedean fields. Indian J. Math. **50**, 125–133 (2008)
3. Natarajan, P.N., Sakthivel, S.: Weighted means for double sequences in non-archimedean fields. Indian J. Math. **48**, 201–220 (2006)
4. Rangachari, M.S., Srinivasan, V.K.: Matrix transformations in non-archimedean fields. Indag. Math. **26**, 422–429 (1964)
5. Katok, S.: p-adic analysis compared with real, Student Mathematical Library, Amer. Math. Soc. **37** (2007)
6. Natarajan, P.N., Rangachari, M.S.: Matrix transformations between sequence spaces over non-archimedean fields. Rev. Roum. Math. Pures Appl. **24**, 615–618 (1979)

7. Natarajan, P.N.: On a scale of summation processes in the p-adic field. Bull. Soc. Math. Belgique **31**, 67–73 (1979)
8. Natarajan, P.N.: Continuous duals of certain sequence spaces and the related matrix transformations over non-archimedean fields. Indian J. Pure Appl. Math. **21**, 82–87 (1990)
9. Natarajan, P.N.: Characterization of some special classes of infinite matrices over non-archimedean fields. Indian J. Math. **34**, 45–51 (1992)
10. Natarajan, P.N.: Matrix transformations between certain sequences spaces over valued fields. Indian J. Math. **39**, 177–182 (1997)
11. Natarajan, P.N.: On the Algebras (c, c) and (l_α, l_α) in Non-archimedean Fields. Lecture Notes in Pure and Applied Mathematics, pp. 225–231. Marcel Dekker, New York (1999)
12. Natarajan, P.N.: Matrix transformations between certain sequence spaces over valued fields II. Indian J. Math. **43**, 353–358 (2001)
13. Natarajan, P.N.: Some Properties of Certain Sequence Spaces Over Non-archimedean Fields, p-adic Functional Analysis. Lecture Notes in Pure and Applied Mathematics, pp. 227–232. Marcel Dekker, New York (2001)
14. Natarajan, P.N.: On the algebra (c_0, c_0) of infinite matrices in non-archimedean fields. Indian J. Math. **45**, 79–87 (2003)
15. Natarajan, P.N.: Some results on certain summability methods in non-archimedean fields. J. Comb. Inf. Syst. Sci. **33**, 151–161 (2008)
16. Natarajan, P.N.: Some more results on the Nörlund and Y methods of summability in non-archimedean fields. J. Comb. Inf. Syst. Sci. **35**, 81–90 (2010)
17. Natarajan, P.N.: More properties of $c_0(p)$ over non-archimedean fields. J. Comb. Inf. Syst. Sci. **38**, 121–127 (2013)
18. Bhaskaran, R., Natarajan, P.N.: The space $c_0(p)$ over valued fields. Rocky Mt. J. Math. **16**, 129–136 (1986)
19. Raghunathan, T.T.: On the space of entire functions over certain non-archimedean fields. Boll. Un. Mat. Ital. **1**, 517–526 (1968)
20. Raghunathan, T.T.: The space of entire functions over certain non-archimedean fields and its dual. Studia Math. **33**, 251–256 (1969)

Index

© Springer India 2015

P.N. Natarajan, *An Introduction to Ultrametric Summability Theory*,
Forum for Interdisciplinary Mathematics 2, DOI 10.1007/978-81-322-2559-1